# CHANGE YOUR GENETIC DESTINY

# Change YOUR Genetic Destiny

## Dr. Peter J. D'Adamo

### with Catherine Whitney

Previously published as *The GenoType Diet*

BROADWAY BOOKS

New York

This book is not intended to take the place of medical advice from a trained medical professional. Readers are advised to consult a physician or other qualified health professional regarding treatment of their medical problems. Neither the publisher nor the author takes any responsibility for any possible consequences from any treatment, action, or application of medicine, herb, or preparation to any person reading or following the information in this book.

Library of Congress Cataloging-in-Publication Data
D'Adamo, Peter.
Change Your Genetic Destiny / Peter D'Adamo with Catherine Whitney.
1. Nutrition—Genetic aspects. 2. Human genetics. I. Whitney, Catherine. II. Title.
QP144.G45D36 2007
612.3—dc22
2007029467
ISBN 978-0-7679-2525-9

PRINTED IN THE UNITED STATES OF AMERICA

*Design by Tina Henderson*

1 3 5 7 9 10 8 6 4 2

First Paperback Edition

*Dedicated to Martha,*
*who kept me strong through the journey.*

"Anatomy is destiny."

—SIGMUND FREUD

# Contents

*Acknowledgments*                                                          xi

*Foreword by Tom Greenfield, ND, DO, MIFHI*                               xiii

*Prologue: Life-Altering Possibilities*                                   xvii

PART I          GenoType: The Key to Understanding
                Who You Are                                                  1

CHAPTER ONE     Your Genetic Autobiography                                   3

CHAPTER TWO     A World of Limitless Potential                              15

CHAPTER THREE   GenoType Whys and Wherefores                                28

PART II         Understanding Your Body's Clues: Why You
                Don't Need a Genetics Lab                                   53

CHAPTER FOUR    Using the GenoType Calculators                              55

CHAPTER FIVE    Strength-Testing Your GenoType                              65

CHAPTER SIX     Ready, Set, Go! Calculating Your GenoType                   85

PART III          The Six Genetic Archetypes:
                  The GenoType Profiles                    107

CHAPTER SEVEN     Meet the GenoTypes                       109

CHAPTER EIGHT     GenoType 1: The Hunter                   116

CHAPTER NINE      GenoType 2: The Gatherer                 127

CHAPTER TEN       GenoType 3: The Teacher                  137

CHAPTER ELEVEN    GenoType 4: The Explorer                 147

CHAPTER TWELVE    GenoType 5: The Warrior                  157

CHAPTER THIRTEEN  GenoType 6: The Nomad                    168

PART IV           The GenoType Diets: Six Individual
                  Roads to Health                          179

CHAPTER FOURTEEN  Getting the Most from Your GenoType Diet  181

CHAPTER FIFTEEN   The Hunter Diet                          191

CHAPTER SIXTEEN   The Gatherer Diet                        205

CHAPTER SEVENTEEN The Teacher Diet                         221

CHAPTER EIGHTEEN  The Explorer Diet                        237

CHAPTER NINETEEN  The Warrior Diet                         253

CHAPTER TWENTY    The Nomad Diet                           269

EPILOGUE          The Future Beyond Tomorrow              286

APPENDIX          The Advanced GenoType Calculator Tables  289

GLOSSARY          GenoType Diet Terms                      299

RESOURCES         Suggested Reading List and Going Further  303
                  Learning More About GenoType            303
                  GenoType Lifestyle Support              306
                  GenoType Practitioner Support           307

INDEX                                                      309

# Acknowledgments

Any work of this nature involves numerous people who've contributed time, talent, understanding, a listening ear, or patience throughout the process. My journey from blood type to genotype has been an extraordinary investigation, and I am indebted to many people who have shared in this process.

First and foremost, I wish to acknowledge and thank Martha Mosko D'Adamo, who provided sage advice about design of this work, as well as serving as confessor and sounding board throughout the project. Martha also took the lead during the copyediting phase of the book, allowing me the continued ability to splice in new material almost until the very last minute.

Heartfelt appreciation to Catherine Whitney, who provided intellectual space and understanding during the early stages when this material was not yet fully formed; Rachel Kranz, who helped with chapter design and programming the flow of material; Chris Fortunato and his team for their attention to detail; Paul and Laura Mittman, from the Institute for Human Individuality (IfHI), for their lavish support of my work, friendship, and goodwill; Dr. Tom Greenfield for beta-testing so many of my more nascent ideas; Ani Hawkinson, proofreader extraordinaire; all the folks at North American Pharmacal, Inc., especially Javier Caceres, Carol Agostino, and Ann Quasarano; the moderators and helpful souls on www.dadamo.com; my co-editors at *The Individualist*; all the patients at the New England Center for Personalized Medicine; Beena Kamlani, Susan Petersen-Kennedy, Denise Silvestro, and all my friends at Penguin Putnam; Amy Hertz, who saw the potential in this material before anyone else; Jenny Frost, the publisher of Crown Books; and my colleagues at Broadway Books, particularly Diane Salvatore, the publisher, and Annie Chagnot, the editor for the trade paperback; super agent and friend Janis Vallely, who kept this entire project from unraveling innumerable times.

Thanks also to my parents, Christl and James D'Adamo, for encouraging me to think differently; my brother, James D'Adamo, and best friend, Robert Messineo, for being such great advisors over the years.

Finally, I send a bouquet of sheer love to Claudia and Emily, who brought tea, snacks, and a ray of sunshine to all those gray, groggy mornings.

# Foreword

*By Tom Greenfield, ND, DO, MIFHI*

Most practitioners who have been using blood groups as part of their treatment approach for some time have learned to recognize the characteristics resulting from a patient's blood group: It is possible to make an educated guess at someone's blood group even before they are tested. There is something about the way an individual appears and behaves that we can recognize as belonging to a category of people who have certain similarities and who respond to the environment in a particular way. Those using the concept of Dr. Peter D'Adamo's Blood Type Diet in their holistic medical practice have also seen patients experience the most profound improvements in health just by making informed food and lifestyle choices. The simplicity of the approach has also empowered people to help themselves, not by using a superficial one-diet-fits-all formula that works only for a few, but by heeding realistic advice that changes according to sound naturopathic principles involving the specific needs of the individual.

It has been twelve years since the publication of *Eat Right for Your Type*, and a lot has happened in that time: The study of the history of humankind has taken a new direction with the work of the Human Genome

Project and its research into the internal workings of our genes. This scientific revolution promised both individualization of medical treatment and opening the door to tracing our ancient ancestry through our genotype. However, genetic testing is still complex, expensive, and beyond the reach of the consumer. Although the theory is there, it will be a long time before genotyping becomes a practical part of prescribing orthodox pharmaceutical products, or genetic genealogy becomes affordable.

There is a wealth of medical research into how we differ from one another, based on clinical observation and easily accessible scientific testing methods. With the advent of DNA analysis, much of this past information has been swept aside in one fell swoop: Political correctness and historical events have made it unfashionable to focus on obvious differences between people. The concept of race has become unpopular and has been declared to have no scientific basis. Resources have been diverted away from measurement of the external signs that distinguish groups of people, to make way for new analytical techniques involving genetic testing, a science that is still in its early infancy. The data from research on human variation that was gathered before we knew how to analyze DNA includes various blood groups, body and head shape, fingerprint patterns, leg length, and many other external signs. This research is still as valid today as when it was first done. Combined, these visible signs tell us more about ourselves than we can know by having an individual genetic test, because some genes tend to stick together in groups or clusters, which have multiple effects on our health. Our environment adds to the picture, changing the way these genes work. We all have visible external manifestations of these genes, something that ancient medicine has been saying in a different way for generations.

Peter D'Adamo is now taking readers one step further: The concept of eating for your blood group has experienced a revolution of its own. What was little more than a footnote in *Eat Right for Your Type* on blood group anthropology is now a book in its own right, telling us even more about ourselves as individuals. Beyond answering the fundamental question "Where do we come from?," *Change Your Genetic Destiny* also informs us about what it means to belong to a particular GenoType and how that knowledge can help us personally to stay healthy. The result of research from this kind of complex science is normally reserved for specialists. However, Peter's analysis is presented to the public in the form of a simple self-discovery manual. The science behind the concept is available for practitioners to study and teach, but the basics are there for everyone to use right out of the book.

It has not been an easy ride, however: Food is an emotive subject at the best of times, and telling people what they can or can't eat according to their genes means any dietary change is not a short-term measure. This has led to controversy over *Eat Right for Your Type*, generated mainly by those who have not understood the concepts behind it or have objected on principle.

Peter D'Adamo has persevered, while allowing people a unique insight into the evolution of his concept via the Internet and through the educational programs of the Institute for Human Individuality. We have observed the workings of an analytical mind belonging to a rare type of person who understands both people and computer programming. It has been a new quest of discovery into why some of us respond differently to healing or behave in certain ways, while staying true to the original concept based on blood groups. The result is a fusion of ancient wisdom, anthropometric techniques from the last century, and modern cutting-edge genetic science, hailing a new era in naturopathic medicine.

On behalf of millions of people whose lives have already been changed forever, and those whose lives will change for the better after reading this book, I would like to thank Peter for bringing this concept to the world.

*June 2007*
*Canterbury, Kent,*
*United Kingdom.*

# Prologue:
# Life-Altering Possibilities

As a naturopathic physician and a researcher, I've always had a deep faith in our ability to take control of our bodies and our lives. Every day, I treated patients who discovered a health, vitality, and joy in life that they'd never thought possible, simply by altering what they ate, which supplements they took, and how they exercised. Since my first book, *Eat Right for Your Type,* was published twelve years ago, I've been privileged to hear the most heartening stories about people whose lives have been transformed by finding the diet and exercise plan that fit them best.

Yet I'll admit I had one blind spot that I'd imagine a lot of you share. I always assumed that the genetic part of our story had already been written. The genes we inherited from our parents, I believed, were the cards we were dealt. I knew we had a lot of leeway in how to play those cards—and I went on to write numerous books that helped a lot of people play them better. But I was pretty sure that the cards we received at the moment of conception were the ones we had for life.

Imagine my delight, then, as I began to discover that we have an enormous power to improve our lives, even when it comes to our genes. True, we can't do anything about which chromosomes we got from our parents;

we can't add new genes to the mix or eradicate old ones. But the genes we get at conception are only the beginning of the story. From our time in the womb through our childhood, youth, and adulthood, *we have the capacity to turn up the volume on some genes and silence others, vastly improving our capacity for health and happiness.* We can understand the trajectory of our life and health—what physical challenges we're likely to face, what disorders we're most prone to—and we can respond effectively. Best of all, we don't need lab tests, drugs, surgery, or medical intervention to perform this miraculous feat. All we need is an understanding of the diet and exercise plan that is right for our particular GenoType—the unique way in which our genes and our cells interact.

Although in traditional scientific practice the word *genotype* is used only to describe a person's actual assortment of genes, I've chosen to use the term in a unique way because I believe that we've sold ourselves a bit short with this narrow, linear definition. The standard definition refers only to chromosomes, but my own use of the word *genotype* includes your relationship to the environment, the influence of your family history from very recent times, and effects of your fetal, or prenatal, history.

An archetype is an idealized model of a person or a concept. Archetypes are timeless and all around us. For example, if we were to think of the archetypal "hero's sidekick," perhaps Robin Hood's friend Little John or Han Solo's copilot Chewbacca would come to mind. Each character comes from a specific time in history, but both fit the timeless archetype.

So in our little universe, words and concept come together to make something new and completely different: a Genetic Archetype—a GenoType.

I developed the concept of the GenoType and identified six GenoTypes by doing statistical analyses of how genes, disorders, and physical traits are known to cluster together. I based my work on the known associations between such traits as, say, fingerprint patterns and particular disorders, leg length and risk for prostate cancer, and tooth shoveling and ancestral diet. Statistical analyses of previously known associations produced six distinct and durable categories that I have labeled the six GenoTypes. The reader who wishes to learn more about the scientific rationale behind the GenoType diet is invited to visit www.genotypediet.com to learn more.

# You and Your Genes:
# A Dynamic Partnership

This book is based on a simple but radical and surprising notion: *We have the power to alter our genes' behavior.* Some of that power belongs to our mothers, during the nine months that they carry us before birth. But a great deal of that power becomes ours as soon as we emerge.

Whether you realize it or not, you've spent an entire lifetime altering your genetic activity. When you took your first sip of wine or beer, you turned up the volume on your body's genetic ability to detoxify alcohol. When you get that fabulous summer tan, you activate the genes that control melanin production. (Melanin is the pigment that protects your skin from the sun by making it darker.) When you pick up an infection, even one as mild as a cold or flu, you boost the activity among your bone-marrow genes, which produce the white blood cells you need to get better. Your genes are not a fixed set of preprogrammed instructions. They are a dynamic, active part of your life, responding every single day to your environment, your history, and your diet.

## The Genetic Town Meeting

Many of us are used to thinking of our genes as dictators, cellular tyrants that insist on doing things one way and one way only. This belief has been supported by several decades' worth of news stories trumpeting discoveries about the role of genes in determining health, immune function, and vulnerability to particular conditions. We've read about the breast-cancer gene, the inheritance of depression, and new tests for such "genetic" diseases as Huntington's chorea and Tay-Sachs. Some news stories have even suggested that genes are responsible for personality traits, like shyness or sensitivity. You could easily get the impression that your genes are your destiny, implacable tyrants who brook no appeal.

In fact, your genes, your body, and your environment all operate together, less like a dictatorship and more like a lively town meeting.

There are the officials on stage whose job it is to announce the problems that need to be solved. There are the loud, abrasive, attention-getting folks who are always rushing to the microphone, making long speeches and insisting on getting their way. And there's the multitude of quiet, solid citizens sitting in the audience, watching and waiting.

You know the town meeting is working fine when you're feeling healthy and vital. Your body is at its optimal weight, you have plenty of energy, and you're fighting off the colds, flu, and more serious infections that occasionally assail you.

But what about those other times, when you're overweight, bloated, struggling with low energy, lackluster skin, and lifeless hair? What about that winter when you seemed to get one cold after another? What about your doctor's warning that you may be a candidate for heart disease, diabetes, or some other devastating condition?

Then you need to shake that meeting up a bit. You might need to keep your inflammatory genes from rushing to the mike, even as you encourage your calmer, anti-inflammatory genes to have their say. Perhaps you want to activate the healing genes that produce more white blood cells to fight infections. Or maybe you need to silence the "thrifty" genes that insist on hoarding every calorie you ingest. *They* think they're protecting you from next year's famine, and they'd love to speak up and tell you so. But *you* know they're causing you to gain weight and setting you up for diabetes, so you really need them to back off.

Every one of us has a huge repertoire of possible responses encoded in our genetic heritage. Our goal is to get some parts of that heritage to speak up while asking others to sit down quietly in the back of the room. This book will tell you how.

As you've probably guessed, your best strategy is diet and exercise. Eating and exercising the right way will keep the right genes talking while gently encouraging the wrong ones to quiet down. But because you've got a unique genetic inheritance—a unique set of characters showing up at *your* town meeting—you need the diet and exercise plans that are right for *you*. The strategies that work for your spouse, your friend, or even your parents might be harmful for you, just as the dietary choices that slim you down and boost your health might put on the pounds and drain the vitality from your loved ones. One-size-fits-all fashions are very rarely flattering to everybody, and the same is true of one-size-fits-all diets. In order to achieve optimal functioning of *your* town meeting, you need to eat right for your GenoType.

## Genetic Medicine and Cellular Repair

So who's at this genetic town meeting and how do decisions get made there? I'll answer that question more fully in Part I of this book, but let me give you a quick answer right now.

One presence at the town meeting, obviously, is the genes you were born with. Those genes determine a whole host of factors, including how you respond to environmental threats, whether you tend toward allergies and asthma attacks, and whether you're likely to store calories as fat or burn them up quickly. These are your town citizens—but they're not the only ones at the meeting.

Another key presence is your environment. You might almost think of the environment as one of the town officials—the one who sets the agenda. Depending on what kind of threat your environment offers, your gene-citizens are likely to respond very differently.

To take one very simple example, a sunny environment puts a new demand before the meeting: *Protect us all from getting burnt!* If you've been born with naturally pale skin, you experience a bright, hot sun as an immediate threat. If your genes do nothing, that sun could easily give you a second-degree burn.

Luckily, you've got a genetic option: Ask your skin cells to produce some extra melanin, which will turn your skin a nice, protective brown. So when the environment announces this new threat, it also sets the agenda—and your genes respond accordingly. The melanin-producing genes come up to the mike and have their say. They keep talking as long as the threat—the sun—is present. When the sun shifts to a less dangerous angle, the environment announces a new agenda: *Get more vitamin D from the sunlight.* Since that goal is more easily accomplished with pale skin, your melanin genes quiet down. Your "pale-skin" genes return to the microphone and your skin reverts to its former color.

Of course, some of us don't have many melanin-producing genes. We're the ones who rarely tan and always burn. No matter how loudly our melanin-producing genes speak up, they can't ever be heard very well. Others of us have naturally dark skin—our melanin-producing genes speak up all the time. So our town meeting can't do *anything* it chooses. But often, it has a certain amount of flexibility. ***To protect us from environmental threats, our genes can often choose to speak up or be silent.***

Another presence at the meeting is your diet. Think of it as another official who sits up on the platform and encourages different responses from the genes in the audience. Once again, we don't have unlimited power to choose a response—we've got to operate with the characters we were born with. But what that diet says from the podium can have a big effect on which genes speak up and which are silent.

For example, some of us were born with "thrifty" genetics—genes whose function is to hold on to every extra calorie and store it as fat. We want those thrifty genes to speak up loudly when food is scarce—in fact, that's probably why we developed them, to protect ourselves from famine. But when food is plentiful, it would be nice to silence them so we don't gain more weight than our body needs. Eating a diet high in carbohydrates and simple sugars invites those genes to hog their places at the microphone, speaking so loudly that no other voices can be heard. Eating more lean protein and enhancing the efficiency of the body's response to a hormone called insulin encourages those "thrifty" genes to step back and lower their voices, so that we lose weight and keep it off.

Likewise, some of us were born with "reactive" genes that instruct our immune system to jump into high gear at the slightest possible threat. These genes induce inflammation—heat, swelling, pain, and extra white blood cells—at the least provocation. They can cause allergies, asthma attacks, rheumatoid arthritis, and a host of other conditions. A diet high in "reactive proteins" such as glutens (found in many common grains such as wheat, rye, and barley) or lectins (found in certain grains, seeds, nuts, and vegetables) invites these inflammatory genes to hog the mike. A diet that minimizes our intake of these reactive foods encourages them to sit down and stop yelling. Our asthma becomes more manageable, our joint pain eases, and we can now tolerate conditions that once had us gasping or writhing in pain.

So our first goal is to get that town meeting back on track. We want to be sure we're hearing from the right parts of our genetic makeup and silencing the genes that are not so helpful. The genes that predispose us to certain diseases and disorders should be encouraged to stay away from the microphone. The genes that help us feel healthy and happy, that lead to a long life and a vital old age, should be invited to have their say.

This is what I call *genetic medicine*, and it is the first aspect of the GenoType program because it's literally designed to reprogram your genes' responses. Accordingly, I've developed individualized diet plans for each GenoType, emphasizing the foods and supplements that will work best to silence some genes and turn up the volume on others, creating optimal health, weight, and vitality. Since every one of your body's functions ultimately begins with your genes, genetic medicine is the deepest and longest-lasting force for change. It's the fastest way to get those positive changes under way and the surest way to hold on to your gains.

Now, what about all the damage that was done when our town meeting was off-kilter? What about the joint pain caused by our reactive genes' overreaction, or the aging of our tissues that resulted from our thrifty genes' retention of all that sugar? Luckily, diet can help reverse those effects as well. That's the second component of the GenoType program, what I call *cellular medicine*. Foods and supplements that nourish and cleanse your cells, ensure the health of your organs, or restore the cell's ability to respond properly to the body's hormone messages are cellular medicines. Through cellular medicine, you can begin to repair the damage that's accumulated through years of poor behavior at those genetic town meetings.

Now, here's the good news. Once you've reprogrammed your genes to behave in healthier ways and repaired the cellular damage that's accumulated, you can ease up on some dietary restrictions and enjoy a more varied set of food choices. You may not be able to eat unlimited quantities of everything—our bodies just aren't built that way. But you probably will be able to broaden your food choices while maintaining your weight, your vitality, and your health. Even better, you'll know how to use food to protect yourself when illness threatens or when stress puts you at greater risk. At such times, you can go back to a stricter version of your GenoType Diet, and then, when the threat has passed, relax your eating habits once again.

## What Is a GenoType?

As you can see, we're interested in both our genes and their response to the environment. We're especially interested in their response to those crucial nine months in the womb. That whole package—our genetic material and its prenatal response—is called a **GenoType.**

Your GenoType determines such seemingly trivial details as the shape of your teeth, the length of your legs, and the pattern of your fingerprints. More significantly, it decides which foods will help you lose weight and achieve vitality—and which illnesses you're most at risk for.

I've identified six GenoTypes: the Hunter, the Gatherer, the Teacher, the Explorer, the Warrior, and the Nomad. These GenoTypes probably developed over the last 100,000 years of human history. They're older than ethnicity and they don't necessarily line up with your family patterns. In other words, you could be the only Hunter in a family full of Gatherers, or the only Nomad in a family full of Warriors. Yes, your GenoType is the product of the genes you inherited from your family, but it's also the result of your prenatal experience. Your genes and your first nine months in the womb combine to create your GenoType.

When you know which GenoType you belong to—which you can learn by taking the simple tests outlined in Part II of this book—you'll know which diet, exercise plan, and supplements can give you the best chance at your optimal weight and maximum health. You'll also learn which diet and exercise can help you prevent the conditions to which you're most vulnerable.

# GenoType and Blood Type:
# What's the Difference?

By this point, those of you who are familiar with my Blood Type Diet may be asking how it relates to the GenoType Diet. After all, both diets are individualized health plans that are geared to your specific biology and your unique health needs. And both depend to some extent on your genetic heritage.

That's true, and I don't want those of you who were good enough to buy my Blood Type Diet books to throw them away! I can assure you, they're still accurate. But in the twelve years since I published the first of these books, I've done further research and learned a lot more about how our bodies work. I can now see that GenoTypes are an even more refined, complete, and accurate way of understanding the human body than blood type alone.

Of course, if you know your blood type, you can include that information when determining your GenoType.

Your blood type is determined by only one gene—one gene out of about 30,000! That single marker gives us an extraordinary amount of useful information, as the millions of Blood Type Dieters can attest.

But GenoTypes reflect the activity of many other genes, not just the one that determines blood type. While blood type is an important aspect of GenoType, it isn't the only gene that matters.

Also, unlike blood types, GenoTypes reflect both genetics and *epigenetics*—the interaction between your genes and the environment. This wider base of information allows us to come up with more detailed and flexible diets that are even better geared to your individual biology and temperament.

I'll share the science behind GenoTypes in Part I of this book. But for now, let me share with you the exciting news that GenoTypes enable me to offer you a far wider set of food choices. For example, in my previous books, I stressed that people with blood type A should generally avoid red meat. That was true for so many blood type A's that I felt justified in saying that. But now I can see that some blood type A readers may have a GenoType—called the "Explorer"—that can be benefited by eating some red meat. On the other hand, blood type A's whose GenoType is the "Teacher" or the "Warrior" should probably continue avoiding red meat, just as I suggested before. Working with six GenoTypes rather than four blood types has allowed me to be more specific—and more flexible.

Also, the whole notion of a Blood Type Diet implied that you had to stick with the same diet for life. After all, your blood type was never going to change! Well, your GenoType won't change either, but since it reflects not just your genes but also your genes' responses to diet and environment, it allows for more flexibility. As you get better control of your town meeting—as you encourage the right genes to speak and the wrong ones to quiet down—you can run the meeting a bit more loosely. As soon as the wrong genes are used to sitting quietly in their seats, you can bring in some dietary choices that you couldn't tolerate before. When the accumulated damage to your cells has been reversed, you can afford to eat a few more things that might have caused problems earlier.

## A Revolutionary New Road to Health and Weight Loss

Since I wrote my first diet book, I've had occasion to read a few others, and I'm struck by how often they promote a one-size-fits-all diet. "Everybody" should eat protein and avoid starch, they announce, or "everybody" should eat more complex carbohydrates and avoid all animal fats.

It's not that these approaches are ineffective. It's that they don't hold true for everybody. The classical physicians seemed to sense a need to personalize their treatments. Both traditional Chinese medicine and Ayurveda, the ancient healing system of India, understand that people aren't all the same. Both approaches tailor nutritional guidelines according to various physical, mental, and emotional differences among individuals. Our own medical system tends to look for across-the-board solutions. If you've got a headache, you take an aspirin. If you've got an infection, you take an antibiotic. And if you want to lose weight, cut out the grains or increase your intake of protein. This approach has worked wonders in treating many illnesses. But it hasn't done so well at creating health, vitality, and well-being. It's not necessarily wrong—but it's definitely not enough.

With the identification of these six GenoTypes, I've created the basis for a scientific understanding of our individual needs for food, exercise, and supplements. Through identifying your GenoType and following your diet and exercise plan, you can achieve and maintain your optimal wellness, vitality, and weight.

The great thing about the GenoType Diet is that it addresses your individual needs so effectively. If your major concern is weight loss, you should know that following the GenoType Diet is the fastest, most effective, and longest-lasting way to lose weight. Each GenoType has different issues that come up around weight and metabolism, which is why each GenoType needs its own diet. Following *your* GenoType Diet and Exercise Plan will give your body all the support it needs to attain and maintain your optimal weight—virtually without effort.

This book will tell you everything you need to know to understand and follow the GenoType Diet. In Part I, you'll learn more about the exciting new science that enabled me to discover the six GenoTypes. In Part II, you'll find out how to determine your own GenoType. Part III will fill you in on the strengths and weaknesses of each GenoType, including health issues, weight concerns, and personality. Finally, in Part IV, you'll find your own, individualized GenoType Diet.

So let's get started! I'm honored to have made this important discovery—and even more excited about sharing it with you.

CHANGE YOUR GENETIC DESTINY

# PART I

# GenoType

## The Key to Understanding
## Who You Are

# CHAPTER ONE

# Your Genetic Autobiography

L et's go back to that town meeting and its lively debates among your genes, your diet, and your environment. We know where your diet and your environment came from. But what about those genes? How did they end up at your town meeting?

When you're born, you inherit a single set of genes—but you get two versions of each, called *alleles,* one set from your father and one from your mother. So right from the beginning, as you are conceived, some genes dominate the conversation while others politely step back.

To some extent, it's a mystery how this works. That's why two parents can have several children, each of them completely different from one another and all of them different from their parents. Who gets Mom's curly hair or Dad's big nose—or Grandma's green eyes or Great-Grandpa's red hair—is to some extent the luck of the draw.

But there are some rules that organize genetic distribution. Some alleles are *dominant,* which means that all other things being equal, they're most likely to prevail. Others are *recessive:* They need to meet up with other alleles just like them in order to express themselves. For example, brown eyes are produced by dominant brown-eye alleles, so when my blue-eyed father married my brown-eyed mother, I was the brown-eyed child who resulted. However, my mother had inherited a blue-eyed gene from her mother, a gene that she could pass on even if she didn't express it herself. There was always a one-in-four chance that my mother's unexpressed blue-eyed gene would meet up with my father's matching gene, and that's why my brother has blue eyes.

Still, brown-eyed genes have three chances out of four to prevail. Why do those genes "speak louder" than the blue-eyed ones? The answer lies in a process called *methylation,* whereby sections of the allele for blue-eyed genes are coated by molecules called methyl groups, causing the section of DNA on which they reside to wind up tightly. Winding up DNA prevents it from being read, essentially turning it off. Methylation is one of the ways nature silences the messages of certain genes. This coating process blocks the DNA at the site of that gene from being "read," and since reading DNA is critical to the gene's ability to express itself, the gene is silenced.

Major amounts of methylation happen at the moment of conception, at about eight weeks into the embryo's development and about a month before birth. After that, it's pretty much downhill; as we age, our DNA gradually loses methyl groups. So far, we can't control what happens at conception: We don't yet know how to make sure that my eyes are brown instead of blue. But research has shown that we can encourage the proper methylation of certain genes *after* conception, primarily through diet. The right diet can help remethylate all sorts of genes that we want to keep away from the microphone, including the thrifty genes that program our cells to store every single extra calorie as fat, or the reactive genes that cause us to have an asthma attack in a dusty room, or the tolerant genes that make us susceptible to one cold after another. (You'll learn how to use diet to your maximum advantage in Part IV of this book.)

But we're getting ahead of ourselves. We're still talking about that mysterious moment of conception, when you get a certain set of genes that promise to express themselves in your unique self. Those are your givens. Those are the genes that are going to come to your town meeting for the rest of your life.

## You and Your Prenatal Environment

At conception, you become a fetus, with your very own set of genes. Your town meeting is populated and you're ready to spend the next nine months in the womb, interacting with your prenatal environment—amniotic fluid, placenta, and a host of other influences. You'll also respond to your prenatal "diet"—the nutrition you take in through what your mother eats and makes available to you. So right from the beginning, your town meeting is under way, with diet, environment, and genes beginning their lively debate.

This debate has enormous significance for the being who will emerge into the world nine months later. For example, your genetic potential at conception encoded a certain range of arterial flexibility. But how strong will your arteries be, and how flexible? A lot of the answers to those questions depend on what happens in the womb.

So that's one of your first town meetings. The genes that encode the delicate branching architecture of your nerves and arteries show up, ready to take part. And there are our friends, environment and diet, up there on the platform as they will be throughout your life, helping your "cardiovascular genes" and "nerve genes" decide what to do. A good diet with lots of protein and healthy fats will encourage your artery genes to make your arteries rugged and elastic. On the other hand, if your mother is unlucky enough to be living in a famine, your "artery genes" will have to compete for those scarce nutrients with the genes that control the growth of other organs and tissues. The result might be a greater tendency for heart disease or high blood pressure. As you can see, the same genes are always at the meeting, but they're responding very differently to the information given by diet and environment.

Obviously, the health of your arteries is only one of the ways that your tissues and organs are affected by your time in the womb. Those crucial nine months help determine whether you gain weight at the drop of a hat or lose weight far too easily. They help nudge you toward a hair-trigger immune response that views the entire world as its enemy, or toward a welcoming immune response that may not always know which invaders to turn away. They predispose you to certain foods that you'll be able to digest easily and turn you away from others that won't suit your particular metabolism and digestive tract. Right from the beginning, your genes interact with diet (in this case, Mom's) and environment (in this case, the womb) to determine who you are.

And then you're born. This is the point at which your GenoType is determined. Your GenoType represents your survival strategy, the decisions that have been reached collectively among your genetic potential, your prenatal diet, and your environment. Although the town meeting will continue for the rest of your life, with genes getting louder and softer in response to diet and environment, certain elements of this meeting are now fixed. They've formed certain patterns—one of six patterns, to be exact, which I've identified as the six human GenoTypes.

## GenoTypes: A Human Survival Strategy

So far, we've been telling the story from an individual point of view— yours, to be exact. But since everyone in the world falls into one of those six GenoTypes, let's step back for a moment and see where these Geno- Types came from.

In the beginning was the environment—a challenging place for our ancestors, to be sure. People had to make sure they could get enough food, that they could survive whatever climate they were born in or migrated to, and that they could resist infections from microbes, bacteria, and viruses.

Genetic inheritance played a crucial role in this survival. People with helpful genes survived; people with less helpful genes died. You've probably heard of this particular aspect of evolution as "survival of the fittest."

Actually, survival of the fittest is more a fantasy than a reality. If only we humans *were* the fittest possible examples of our species, we'd all be a lot healthier, and I might be out of a job! In fact, evolution is more like a game of chance than any true contest. Sometimes the good players win; sometimes it's just the lucky ones who survive. Sure, a lot of our genetic inheritance helps us beat the odds, but there's also a big chunk of it that gets in our way or doesn't play any useful role at all:

**1. Sometimes the "good genes" that help us survive also have their downside.** The genes that instruct our immune systems to react swiftly to bacterial invasion also overreact to produce allergies, asthma, and rheumatoid arthritis. The genes that instruct our fat cells to hang on to every calorie in order to survive a famine also contribute to obesity and diabetes. The genes that program our immune systems to respond calmly to the environment, enabling us to tolerate a wide set of circumstances without getting sick, also tell that immune system to tolerate some deadly invaders that it ought to repel.

**2. Sometimes mutations simply "happen."** Our genes are designed to reproduce themselves in exact copies, but of course, that's not always how it works. Sometimes a gene reproduces slightly differently and that variation becomes part of our genetic heritage. Skin color probably emerged this way, as a mutation of the gene that determines skin pigment. Although originally we all had dark skin, those of us living in northern climates were better able to absorb vitamin D from the scarce sunlight if our skin was lighter; we also had less need of the darker pigment's protective function. So having lighter skin was a mutation that survived because it actually did contribute to our survival.

To some extent, though, these mutations were just random, such as the mutations that produced such diseases as Huntington's chorea or Tay-Sachs. Or, sometimes, they represent a trade-off between the lesser of two evils. For example, having sickle-cell anemia seems to be somewhat protective against malaria, so people with the genetic tendency to have one disease are meanwhile being protected against another.

As you can see, some mutations make our lives better, some make them worse, and some probably don't make much difference either way. But the "bad" or "neutral" mutations don't necessarily die out, and as we'll see in a moment, the good ones don't necessarily survive.

**3. Sometimes who survives is just the luck of the draw.** If all the strong young men die in a huge battle, the male survivors are not necessarily the healthiest—but they will have lived the longest. If an avalanche destroys three-fourths of a village, the remaining one-fourth who stagger down the mountain may be more lucky than fit. There is also what scientists call the "founder effect": When a small group splits off from a larger group and migrates to a distant land, its members may carry only a fraction of the original population's genetic potential. Whatever genes they managed to take away from the larger group, those are the ones that survive—and they're not necessarily "the fittest." Nevertheless, these are the survivors, the ones who pass their genetic inheritance down to the rest of us.

Why should you care about any of this? Because the final effect of this whole process has been our six GenoTypes, which are extremely useful but very imperfect strategies for survival. Every GenoType has its upside and its downside. I personally can imagine ways to improve every single one of them, and once you get to know them better, I'm sure you will, too. In fact, that's the point of this book: What the GenoTypes have begun, we can complete, maximizing their strengths and minimizing their weaknesses through diet, supplements, and exercise.

Remember, your genes don't stay fixed in their tracks. Instead, they keep reshaping themselves and you, just as they did when you were in the womb. Your cells are constantly reading the environment they're in and altering their functions accordingly: Toxic or safe? Food-rich or barren? Threatening or welcoming? These conditions prompt your cells to turn various genes on and off, depending on how the environment is affecting them. These instructions are implemented at the town meeting, where the volume is turned up on some genes and turned down on others.

The end result is our six GenoTypes, each of which has its own unique pattern of "noisy" and "silent" genes. Accordingly, each of our six GenoType Diets is designed to alter that pattern to promote your optimal health and weight.

## Meet the GenoTypes

So let's take a closer look at these GenoTypes. What possibilities for human survival are encoded within these genetic and prenatal patterns?

Before I introduce you to the GenoTypes, I want to caution you against two common mistakes. First, these GenoTypes do not correlate in any way to ethnic patterns. They seem to have developed tens of thousands of years before ethnicity emerged, and with the exception of the Nomad GenoType, which seems to have more than its fair share of redheads, they don't follow any of the statistical patterns that do correlate with ethnicity, skin color, eye color, hair texture, hair color, ancestry information markers, mitochondrial DNA, Y-chromosome DNA, and a host of others with far more technical names. In the end, I could find only a few weak links with some of the ancestral DNA markers but absolutely no correlation with ethnicity—none. Just as anyone can be hot-tempered or optimistic, anyone can be any one of these GenoTypes, no matter their racial or ethnic identity or who their ancestors are. (And, just to make this crystal clear, no one can use this theory to "prove" that any ethnic group is superior to any other!)

Second, although I think it's useful to speculate on how these GenoTypes emerged, I don't really have the archaeological or anthropological evidence to tell that story. I know that six GenoTypes exist now, and I can deduce how they developed, but that's the part of the story for which the evidence is still largely circumstantial. That's because I've worked backward—I've identified the six GenoTypes that currently exist and then speculated about how they got here. Someone else will have to fill in the rest. Meanwhile, all *you* need to know is that by following the correct diet for your GenoType, you can multiply exponentially your chances for health, vitality, and maintaining your ideal weight.

So let's meet the GenoTypes! You'll read about each of them in depth in Part III, but here's a preview that makes it clear how each GenoType represents a unique survival strategy. I like to think of them as easily recognizable archetypes, striding over the challenging terrain of the Paleolithic and Neolithic planet. Each GenoType has unique strengths that give them an advantage when dealing with food shortages, climate change, and infectious diseases, and each of them has unique weaknesses as well.

**GenoType 1: The Hunter.** This is one of the earliest surviving responses to the challenge of human survival. Beleaguered by what must have seemed like an overwhelming set of environmental challenges—hunger, climate, infectious diseases—Hunters exemplify a *reactive* approach, a quick and powerful response to every potential threat. "Shoot first and ask questions later" might be their motto. The upside: They've got immune systems that act swiftly to attack microbes, viruses, and bacteria that try to kill them, and they're terrific at metabolizing the meat that is their primary source of nutrition. The downside: Their hair-trigger immune response can sometimes lead to overreaction in the form of allergies, asthma attacks, and other inflammatory conditions. Sometimes, as with the viral gastroenteritis to which they're also prone, their immune response may target their own tissue as part of its effort to repel the invaders, a kind of "friendly fire" that can create more health problems than it solves. Another downside is Hunters' inability to digest grains and some other types of food—it's just not what their system is geared for. So our GenoType 1 Diet for Hunters helps calm their immune systems and keeps them away from the kinds of "reactive proteins"—lectins and glutens—that are likely to set them off.

**GenoType 2: The Gatherer.** If you need to survive a famine—and many of our ancestors did—GenoType 2 is designed to get you through it. Gatherers have *thrifty* genes whose primary goal is to hang on to every ingested calorie for dear life—literally. If the Hunter's motto is "Shoot first and ask questions later," the Gatherer lives by "Whoever dies with the most wins." Gatherers learned in the womb that there wouldn't be much food when they got out, so their town meeting quickly worked out a system whereby food conservation was its top priority. The upside to this approach is obvious: It kept them alive and able to bear or father children. The downside is equally obvious: In a more affluent society, Gatherers tend toward obesity and diabetes. Many of my Gatherer patients assure me that they've cut down their caloric intake to an almost starvation level—and yet they still can't lose weight. No wonder: The more they starve themselves, the more their town meetings insist on hoarding food! Our GenoType 2 Diet helps Gatherers rev up that sluggish metabolism and return to their ideal weight, reducing the risk of diabetes and reversing the negative consequences of obesity.

**GenoType 3: The Teacher.** The Teacher represents the third basic response to a challenging world: *altruism*. "All you need is love" is the Teachers' motto, and their immune systems reflect it. Perhaps because this Geno-Type emerged during a time when people were migrating more and living in more varied environments, Teachers are able to tolerate a wide variety of unfamiliar bacteria, viruses, and microbes, avoiding the hair-trigger symptoms that plague the Hunter. Unfortunately, they sometimes welcome infectious elements that they would do better to repel. Teachers may live for a long time without symptoms and then discover that digestive problems, lung disorders, or even cancer have been building within them for years. Our goal for Teachers is to protect their stomachs, colons, and lungs from the wear and tear of the environment. We want to keep their "good bacteria" happy and numerous so they can crowd out the disease-causing "bad" bacteria, yeast, and viruses. We also want to make their immune defenses more efficient by "teaching" them with food, supplements, and lifestyle to be more protective and discriminating.

**GenoType 4: The Explorer.** The first three GenoTypes were probably developed by our ancestors some 50,000 to 75,000 years ago. The Explorer is a newer model, maybe about 20,000 to 30,000 years old. Like Hunters, they're reactive, but unlike Hunters, they're highly idiosyncratic, reacting intensely to some environmental threats and not at all to others. For example, if you're one of those folks who can't drink coffee because it keeps you up all night, you're almost certainly an Explorer. I have no idea whether Frank Sinatra was an Explorer, but he did popularize their motto, which is "I did it my way." My theory is that in every population, you need some folks who do things very differently, so that if the majority turns out to be wrong, someone is there to offer an alternative. Explorers can do that—but sometimes they just seem naturally contrary. They're more often left-handed, Rh-negative, and asymmetrical: Their left and right sides don't match, right down to their fingerprints. They seem to have their own ways of digesting food, responding to disease, and otherwise coping with an ever-changing, unpredictable environment. I think of them as glacial refugees, continually forced from one home to another as they tried to escape the ice. They never had the chance to settle down into a stable relationship with one environment, so they couldn't afford the blanket reactivity of the Hunter. Accordingly, they fine-tuned their responses—but in unpredictable and sometimes inexplicable ways. In modern times, they seem to have emerged from a highly unstable prenatal environment (asymmetry is usually a sign of that), and they've figured out while still in the womb that they're going to have to adapt to wildly changing conditions. Their upside is that they're good at that. Their downside is that odd things—like caffeine—can keep them up for hours. Our dietary goal for them is to develop calmer, more stable responses to the world, protecting them from their idiosyncratic but very vulnerable Achilles' heels.

**GenoType 5: The Warrior.** Like the Gatherer, Warriors are a Thrifty Geno-Type, but I believe they, too, are a newer GenoType, perhaps dating from the Neolithic Revolution (about 11,000 B.C.E.) to the Iron Age (starting around 2000 B.C.E.). Instead of displaying the overwhelming thriftiness of the Gatherers, who hoard pretty much any calorie they can get their hands on, Warriors are more selective. If they're physically active, their metabolism burns hot; when they lead a sedentary life, they tend to put on the pounds with alarming speed. I've called them "Warriors," but perhaps "war survivors" would have been a better name: I've often thought that this GenoType developed in response to the scarcity of the post-Neolithic period, when agricultural technology was still at a very low level, trade was limited, and there were enormous disruptions resulting from war and the spread of "civilization." Earlier in our history, most disasters were natural; then we started creating human-made catastrophes—war, conquest—and I think the Warriors were among the first survivors of these. The survivors in these warrior societies often had to produce large numbers of children, who had an urgent need to progress rapidly to adulthood. In a sense, their genetic inheritance is like the multiple copies of the photocopy machine, getting blurrier and less reliable with each replication. On the other hand, the harsh conditions of this "survivor" life have given Warriors some remarkable strengths: endurance, stamina. "Time flies when you're having fun" might be their motto—but luckily, with the GenoType Diet, we can do a lot to prolong their life, health, and vitality. (I have a vested interest in the matter, since I'm a Warrior myself!)

**GenoType 6: The Nomad.** The Nomad is also a newer GenoType. Their survival strategy reflects a life of travel, encountering different environments and having to cope with a wide variety of challenges. Some of the first people to create a migratory life based upon the use of the horse, Nomads moved quickly over large swaths of territory, passing through a wide variety of climates and terrains. In such a life, any single survival strategy would have only a limited value, and it wouldn't pay to be too reactive to any particular environmental factor. You'd have to learn tolerance—but a limited tolerance, taking in only so much of your environment and keeping your guard up at least some of the time. "A new career in a new town" is the Nomads' motto, and, like the Teachers, their response to the environment is more toward the altruistic, tolerant side. They're a bit more selective than Teachers, though, and somewhat better at filtering out hostile invaders like microbes and bacteria. The price they pay for their more selective immunity is a problematic connection between their immune system, their cardiovascular system, and their nervous system, resulting in a lack of coordination among the three. This makes the Nomad prone to highly idiosyncratic health problems, such as chronic viral infections, debilitating long-term fatigue, and memory problems. Although these are physical problems for sure, they are also often the result of a much deeper disconnect between the Nomad's mind and body. Although well-functioning Nomads usually have wonderful abilities to heal their bodies through visualization, meditation, and relaxation, stress can cause their normal mind-body integration to disconnect so that their physical systems spin out of control. Our goal here is to defend those few places where Nomads' immune systems are susceptible to problems and to increase communication within their bodies.

Now you've got some idea of how the GenoTypes developed and a thumbnail sketch of each GenoType. You may be wondering how you can determine which GenoType you are. You can just take my word for it, fig-ure out your GenoType in Part II, discover your GenoType in Part III, and begin following the diet for your GenoType in Part IV. But if you'd like to know more about what those tests are based on, turn the page and start reading Chapter 2.

# CHAPTER TWO

# A World of Limitless Potential

So far, we've been looking at changes that took place over tens of thousands of years. Now we're going to focus in on much shorter time spans: your life in the womb and your life right now. When you understand just how dramatically your diet and environment can affect your genes, you'll have the power to control your own genes—the power to become the best YOU that you can be.

The first step is for you to understand more about how your genes "listen" to your environment, both in and out of the womb. This is the growing science of epigenetics, the way your genetic activity can be changed by your cells—and then passed down to future generations.

## Genes and Epigenetics

Let's step back a moment and remember where it all starts: with our genes. Now, I'm the first to admit that our genetic code is a miraculous thing—the material that helps to make us human. And when a group of scientists formed the Human Genome Project with the ambitious plan of mapping every one of our genes, I was as excited as anyone. A lot of people thought that the Genome Project was going to discover the key to human existence, the very essence of what makes us who we are. While I didn't go that far, I'll admit that I shared in the general enthusiasm. Identifying all the human genes there were seemed to offer a tremendous potential in understanding health, vitality, and human nature itself.

But then the scientists stumbled upon an unexpected discovery: We just don't have that many genes. In fact, once we started counting, there didn't seem to be nearly enough genes to account for our extraordinarily complex selves. And while we'd all agree that we humans are more complex and varied than any other species, we actually don't have the most genes—not by a long shot.

In fact, we humans have only 30,000 genes. True, that's more than most fungi (around 6,000) and many worms (around 19,000). But it's fewer than many fish (40,000) and most plants (around 60,000). As the Human Genome Project drew to a close in 2001, biologist David Baltimore commented, "Unless the human genome contains a lot of genes which are invisible to our computers, it is clear that we do not gain our undoubted complexity over worms and plants by using more genes."

No? Not more genes? Then what *does* give us our undoubted complexity?

Aha! That's where epigenetics comes in: the science of how our genes respond to our environment, creating differences that we can then pass along to our children. Or, from a more selfish point of view, creating differences that our parents passed along to us. It's as though our genes are the piano keys—but epigenetics is the composer. The melodies and harmonies you can write for those eighty-eight keys are seemingly infinite, even if they can never stray outside the basic range of the piano. That brings us back to GenoTypes—they're the six basic melodies that we humans have come up with as our 30,000 genes interact with the environment. So now we can look at the following formulation:

**30,000 genes + prenatal experience +**
**last 100,000 years on earth = 6 GenoTypes**

I guess it's time for a confession. There probably aren't only six Geno-Types. There are probably 7.5 billion—one for every human currently alive on the planet. Just as each of us is a unique person with a unique appearance, personality, and set of abilities, so does each of us have a unique GenoType—our very own set of interactions between our 30,000 genes, our nine months in the womb, and our life since then. These experiences affect our bodies, our health, our vitality, and our weight. They also affect our genes. Some of them, if we're fertile and reproducing, can be passed on to our children. All of these interactions affect us on a daily basis—and they all continue to affect our genes.

As you can see, it's a dynamic, continuous process that won't stop until you die. If you've had children, it continues even after, for them and their children and their great-great-grandchildren. As a result, "your" self—the interaction between your life experience and your genetic inheritance—is unlike any other on the planet. You, your genes, and your life experience come together to create a unique melody that no one else will ever quite duplicate. I invite you to take a moment to share my wonder at that fact.

Of course, I can't design 7.5 billion individual diet and exercise plans—but fortunately, I don't have to. Remember our example of the "Hero Sidekick" archetype? Now, I doubt many people would think that Little John and Chewbacca stand much chance of being mistaken for the other, but they were clearly "sidekicks" in the minds of anyone who had ever read *Robin Hood* or seen *Star Wars*.

As it happens, the survival strategies we've worked out with our 30,000 human genes, our common prenatal experience, and our 100,000 years on the planet have fallen into six basic patterns. I can't tell you why there are six any more than I can tell you why there are four blood types—it just happened that way. Maybe if we're on the planet another 100,000 years, we'll develop another six, or maybe we'll get rid of some old ones and replace them with new ones—who knows? For now, my statistical analysis has shown that we've got six GenoTypes, each with its unique strengths and weaknesses.

So let's take a closer look at how our cells and our environment interact to produce our GenoTypes. We'll learn a lot about where we've come from. More important, understanding epigenetics—the way our cells and our environment interact to produce differences that we can inherit and pass on—can help us make better choices about where we're going.

## Methylation: Adjusting the Volume Control on Our Genes

Methylation is a very important process in epigenetics, and it's one of the keystones of the GenoType Diets. So let's take a closer look at how methylation helps us turn the volume on our genes up or down.

In Chapter 1, we saw that our genetic heritage is transmitted through alleles—the two sets of genetic possibilities that we inherit from our parents. Some alleles, such as those for brown eyes, tend to dominate others, such as those for blue eyes. When a brown-eyed parent and a blue-eyed parent combine their alleles into the gene for a single embryo, the brown-eyed allele has a four-to-one chance of predominating.

Now, here's where methylation comes in: It is the process that silences the blue-eyed allele. Of course, that allele—the genetic potential for blue eyes—is still carried within the embryo, and when the embryo grows up and has kids of its own, he or she can potentially pass that potential on. Meanwhile, though, the methylation has silenced the allele for blue-eyed genes, allowing the brown eyes to speak.

Methylation affects two aspects of our genetic heritage: the qualities we can't change once they're established (such as eye color, skin color, and blood type) and the qualities we can (such as metabolism, immune function, and general vitality). For example, if the genes for diabetes and obesity are methylated while a person is still in the womb, those genes are "silenced" and the person is far less likely to develop those disorders. Methylation helps determine such key issues as whether your reactive genes overreact to pollen, or whether your thrifty genes hang on to every ounce of body fat, or whether your tolerant genes welcome some invaders that they ought to repel. So it's important for us to understand how and why it works—especially since the GenoType Diet will help you take advantage of the benefits of optimum methylation to achieve your optimal health, vitality, and weight.

How does methylation work? Let's start by remembering that every cell contains a complete replica of our DNA within its nucleus. Sometimes, a methyl group attaches itself to one of those DNA strands. The cell can no longer read the part of the DNA where the methyl group is attached, and so it ignores the genetic command contained in that part of the strand. In effect, one or more of your genes have been silenced.

Of course, each one of us has trillions of cells, all with identical DNA instructions in the nucleus. So if a gene is to be genuinely silenced, the methylation process has to occur in a significant number of those cells. Otherwise, the unwelcome genetic message—*cause diabetes* or *hold on to fat*—can still get through.

Remember, too, that this is a dynamic process: It's happening all the time. Right this very minute, some cells in your body are dying while others are reproducing themselves. Maybe last week, you had a majority of methylated cells and the muffled genes were sitting quietly at the back of your town meeting. But perhaps this week, due to a change in your diet or stress levels, your methyl levels are down and your cancer-causing or fat-retaining genes are back at the microphone. If they stay there long enough, your town meeting may make some very dangerous decisions.

Genes can be methylated in two ways, globally and locally, and knowing this can help you from reading parts of this book and saying to yourself, "Wait a minute. I thought he said that methylation was a good thing. But here it sounds like a bad thing instead." Where genes are methylated locally, it is usually at the beginning of the gene, essentially where the gene tells the DNA machinery, "OK, this is my beginning. Start reading here." One of the most interesting and indeed sinister examples of methylation is when cancer cells use it to silence tumor-suppressing genes—the very genes that are supposed to protect us. This often involves methylating the front of the gene, which turns it off.

Global methylation is the methylation pattern throughout the rest of the gene. In early childhood, our genes start off well-methylated, but over time as we age, we lose more and more of our genetic methylation. This is probably why many diseases develop with age; we lose the ability to control the genes that produce them. As an aside, this is probably why you should be sipping green tea as you read this part of the book. Green tea is one of only a few foods that perform "methylation jujitsu" on our genes: taking methyl groups off the front of the gene (where they almost always cause problems) and remethylating the rest of the gene (where they almost always do some good).

We don't know everything about what causes either the beneficial or the detrimental methylation process to occur. But we do know that diet, supplements, and exercise play a huge role in quieting the genes we most want to silence. One of my goals with the GenoType Diets is to fine-tune exactly the kind of diet and exercise that will best address *your* genetic and epigenetic patterns so that you get exactly what you need to keep silencing the hurtful genes and encouraging the helpful ones.

## Histone Acetylation: Helping Us Read Our Genetic Code

The second major epigenetic process is known as **histone acetylation**. **Histones** are spool-like molecules that cause DNA to wind into dense coils. That's important, because if you stretched out one of your DNA strands, you'd discover that it was six feet long (albeit microscopically thin). Histones are like little spools that allow your DNA to fit inside your cells as it winds around them.

Now, when DNA is all coiled up like that, it can't be read. So any DNA that's coiled around a histone has basically been silenced. It may show up at your town meeting, but it isn't going to speak.

Of course, we need our DNA to speak some of the time; otherwise, our cells couldn't reproduce. Our body gets DNA to unwind by placing molecules called **acetyl group**s onto the histones. These acetyl groups unwind the DNA—and it speaks. Then, when it needs to be wound back up and silenced, enzymes remove the acetyl group and the DNA spools around your histones once again.

Actually, the process is a bit more intricate than that. Whenever your DNA is wound up on its little histones, one small portion of it *is* available for reading, even though most of it is silenced. So if you imagine all the trillions of cells within your body, each with its own DNA, tiny portions of your DNA are available for reading in each cell. The trick, as with methylation, is to try to get the "right" portions of the DNA to be read while keeping the "wrong" parts wound up on their spools.

This process is controlled by two enzymes. One enzyme gets our DNA to unwind, the other gets that same DNA to curl up around its histones. As you've probably guessed by now, the levels of these enzymes are affected by your prenatal experience, environment, diet, and lifestyle. Moreover, many of the same foods and supplements that encourage the right kind of methylation encourage the right kind of histone acetylation as well. So that's another way that your GenoType Diet can affect which of your genes speak and which are silent.

Like genes, histones can also be methylated. As a matter of fact, it is thought that the more permanently silenced alleles (like the blue-eye allele from my dad) are actually silenced by turning off their histones, binding them up for good.

## How We Inherit Epigenetic Patterns

You already know that your genes are inherited. Well, guess what? Your epigenetic programming was inherited, too. At the moment of conception, your parents passed on to you not only their genes but their distinctive styles of epigenetic programming, their own unique patterns of silencing genes or encouraging them to speak. So you start out at conception with inherited patterns of epigenetics—but from the moment you exist, you begin developing your own patterns as well, responding to the conditions in your mother's womb.

For example, if a mother faces famine, malnutrition, migration, war, or other stressors, the fetus within her might also be malnourished and perhaps deprived of oxygen. In fact, with enough prenatal stress, a fetus may be so oxygen-starved that it faces the equivalent of what an adult might endure standing at the top of Mount Everest without an oxygen mask. Likewise, if the mother eats poorly or diets strenuously—before and especially during her pregnancy—the fetus gets even less of what it needs. If Mom smokes, ingests alcohol, or takes drugs, fetal conditions get even worse.

So the time in the womb teaches you an important lesson about what you might expect when you get out into the world. Based on how much oxygen and nourishment is available and on how many prenatal stresses and shocks you receive, your town meeting begins to make some early decisions about which survival strategy will work best once you are born.

For example, you might have learned that calories are scarce and uncertain, so you'll need to turn up your thrifty genes to top volume, conserving every possible calorie as fat against the uncertain future. Or maybe you've learned to hyperreact, mobilizing your immune system against the various shocks and toxins that assail you. Or perhaps your ancestors migrated to a new environment and had to adapt their immune systems to a new local diet. Those old reactive immune systems were not going to be very helpful if all they did was cause you to die of starvation! Whatever your survival strategy, you start learning it as soon as you're conceived, and you spend your first nine months working it out.

## Learning What to Expect
## Before You're Born

Even though you can't go back and change your life in the womb, I want to spend a bit more time on this prenatal period, because it makes clear just how dramatically your genes can be altered by the environment. True, our postnatal experience isn't quite as dramatic, and no diet and exercise plan will ever have the impact on us that our nine months in the womb has had. But diet and exercise can play a big part in altering our genes, especially if they're geared, as the GenoType Diets are, to our existing genetic and epigenetic programming.

So here's one famous example of epigenetic programming triumphing over genes. The Agouti mouse is a special animal developed in laboratories, deliberately bred to be obese, with tendencies to diabetes and other obesity-related disorders. As it happens, Agouti mice also have golden coats.

You might assume that once their genetic programming had been accomplished, Agouti mice would have fat offspring until something else changed their genes. Not so! When Agouti mothers were fed a special diet that contained large amounts of methyl—specifically soy, which is high in the amino acid known as methionine—they had slim offspring with no special tendency to diabetes and, as it happened, brown coats. (The changing coat color is an example of that genetic randomness we talked about in Chapter 1. There's no survival-related reason why the fat mice were golden and the thin mice were brown; the neutral color gene and the harmful obesity gene just seemed to go together.)

A famous photograph shows two Agouti siblings: a fat older mouse with a golden coat and a thin younger mouse with a brown coat, both offspring of the same mother, born only a year apart. Both mice had the Agouti "fat" gene, but the younger mouse's gene had been methylated—silenced—in the womb because of the mother's diet.

Here's another, more tragic example of prenatal epigenetic programming. It's actually the first time scientists began to realize that prenatal experience literally could reprogram our genes. In the final years of World War II, the Allies engaged in an unsuccessful military endeavor in Holland known as Operation Market Garden. Despite the name, the operation actually had nothing to do with either gardens or food—but its aftermath did. When the operation failed, the Nazis blew up all the dikes and dams in western Holland, cutting off the Dutch from any access to food. The winter of 1944–45 became known as the *Hongerwinter*, as 30,000 Dutch starved to death and the survivors were struggling along on official rations of 400 to 800 calories a day. When you compare that to the allowance of 2,000 calories per day for most adults and 2,300 for pregnant or nursing mothers, you get a picture of how extreme the situation was.

Even so, almost 40,000 babies were born during Hongerwinter—and in the 1960s, researchers began to study what had happened to these famine survivors after they grew up. Not surprisingly, fetuses who had been in their last trimesters during the height of the famine had been born with low birth weights. Later, as more food became available, these children grew up normally. But as adults, they suffered from very high rates of diabetes.

Meanwhile, fetuses that had been in the first six months of gestation during the height of the famine were born with normal birth weights—

apparently, they had caught up during their last trimester. However, when they reached adulthood and had children of their own, their babies were unusually small. These prenatal famine sufferers had learned in the womb that food would be scarce—and their genes altered accordingly.

Fetuses who had been exposed to famine during gestation also went on to develop more obstructive pulmonary and kidney disease, including higher rates of atherosclerosis, or clogged arteries. They also had a greater tendency to high blood pressure, higher rates of obesity, and a threefold increase in heart disease. Girls grew up with significantly higher rates of midlife diabetes and obesity; boys suffered from higher rates of schizophrenia and had exaggerated biological responses to stress, such as excess production of stress hormones or racing hearts accompanied by increased rates of hypertension.

Clearly, starvation in the womb had profound effects for the children's later health, even though they'd had access to sufficient food throughout their entire childhoods and adulthoods. Not only did the prenatal experience affect the children later in life, it also affected *their* children—at least, those born to parents who had been in their first trimester when the famine struck.

A theory to explain these effects was suggested by David Barker, a British researcher working at the University of Southampton in the mid-1980s. Barker had studied similar effects in British populations, and he came to the conclusion that a pregnant woman's body could modify the development of her unborn child to increase its chances of survival.

In Barker's work, the focus was on "survival under scarcity"—how to help your child endure a famine. But as we'll see in Part III, I believe that all the GenoTypes represent maternal programming to help their children survive. Some, like GenoType 2, the Gatherer, are geared to defend against famine. Others, like GenoType 1, the Hunter, are intended to protect us from environments full of infectious diseases. Whatever the potential problem, the mother's cells identify it and send appropriate instructions to the fetus's "town meeting." That's how our early patterns are set—and unless we reprogram them through a diet and exercise plan specifically geared to address them, that's how they continue throughout our entire lives.

## Epigenetics and Aging

Epigenetics may also enable us to exert some control over the way we age. As we get older, our cells tend to metabolize less efficiently and our tissues become more likely to accumulate age-related by-products. As a result, our organs don't work as well and we experience a general loss of vitality. If we could protect our cells' ability to metabolize—that is, if we could get our genes to give our cells better instructions as they got older—we might extend our lives and our vitality.

Methylation seems to be a key factor in aging well. Mice, which live for only a few years, methylate badly. Humans, who live into their seventies, eighties, and beyond, methylate well. Clearly, optimal methylation would enable us to live still longer.

In addition, the right kind of histone acetylation can also increase longevity. A famous study showed that mice on calorie-restricted diets—which affected their histone acetylation—lived up to 50 percent longer than ordinary mice. Happily, the GenoType Diet *isn't* calorie-restricted—it's calorie-intelligent. But the study shows how powerful diet can be in encouraging our "longevity genes" to speak.

Another epigenetic issue in aging concerns our *telomeres.* Think of the telomere as the plastic cap that keeps a shoelace from fraying. Telemores sit on the ends of our chromosomes to keep them from fraying so that our chromosomes don't lose genetic information as our cells divide and reproduce.

Remember, our cells are constantly replenishing themselves. Eventually, every cell must divide and create two new cells, each with its own perfect copy of the first cell's DNA. Methylation during that process affects which genes speak and which are silent. But we still want the genes to copy themselves perfectly, even if some of them are muffled. Otherwise, we lose our ability to create new skin, blood, bones, and organs. It's as though everything in our body just wears out and we've lost the ability to replace it. In fact, our cells are able to divide an average of fifty-two times before our telomeres wear out.

Of course, if our telomeres are longer, our genes are better protected and we may have more time, perhaps as much as five years. So one of my goals with the GenoType Diet is to keep your telomeres in good repair—particularly for GenoType 5 Warriors, who appear to have a tendency to age quickly and "wear out" fast.

How do we improve the epigenetic control over our aging process? You've probably already guessed the answer: through a GenoType-specific diet geared to our own particular body chemistry and metabolism. Some GenoTypes will methylate better if they consume more soy, nuts, and seeds. Others need eggs. Still others would do well with cheese. Some GenoTypes need to concentrate on telomere repair, others on keeping their DNA tightly wound around their histones. And because each GenoType often has different weaknesses and strengths, they have different goals as well and can benefit from different supplements. (However, some supplements are good for everyone. I've already mentioned green tea, and it seems that folic acid and vitamin $B_{12}$ can potentially improve methylation in all six GenoTypes.)

Finally, let's look at an aging process known as *glycation*. Glycation is a kind of caramelization process that happens in your body, greatly resembling the caramelizing of an onion or an apple. It occurs when a sugar molecule—such as fructose, from fruit, or glucose, from refined grains—binds to a protein and damages it. The damaged protein interferes with organ function, blood profusion, hormone receptivity, and kidney function, and it might cause cataracts and neuron damage as well.

The protein-plus-sugar molecule is known as an advanced glycation end product, or AGE for short. AGEs produce fiftyfold more toxin-free radicals than nonglycated proteins, and they have the potential to wreak havoc in your body. They've been implicated in atherosclerosis (hardening of the arteries), high blood pressure, diabetes, arthritis, and Alzheimer's. Increased levels of AGE molecules also create inflammation, as well as the universal symptoms of aging: reduced organ function, weakened lungs, compromised blood vessels, general reduction in blood flow, and a loss of collagen under the skin—aka, wrinkles.

I'm sorry to tell you that AGE molecules are fairly hard to get rid of and that our bodies tend to eliminate them very slowly. But the good news is that if you eat right for your GenoType, you can increase the natural ability of your genes to provide antioxidant protection, which will both prevent AGE molecules and help eliminate them.

## Reading Your Prenatal History

Now that you understand why our prenatal history is so important, it's time to find out what happened during those crucial nine months. No, I'm not suggesting that you call your mother. You can learn everything you need to know by working through the next few chapters.

# CHAPTER THREE

# GenoType Whys and Wherefores

**A**t this point, you may be thinking, "I understand why I need to know more about my own genes. After all, if I've got the gene for breast cancer or a genetic tendency to Alzheimer's, I certainly want to know about it. And if your diet can help me muffle those genes, so much the better. But why do I need *your* calculators? Why not just rely upon a genetics lab?"

Well, for one thing, genetic testing that's done in lab tests is expensive and time-consuming. You can only test for one gene at a time, and lab tests exist only for some disorders—not for the kinds of tendencies we've been talking about (weight gain, reactivity, or an inappropriate tolerance of "invaders"). Not only will you get very limited information concerning only one gene at a time, but you might not think to test for conditions you don't yet know you're at risk for.

Second, we don't have lab tests for all the genetic disorders we know about. For example, there's no direct laboratory test for Alzheimer's. One gene, known as APO E4, is positively correlated with Alzheimer's, but at about only a 35 percent rate of prediction. In other words, roughly only one in three people who test positive for this gene will go on to develop the disease later in life. As it happens, taking a thorough family history will give you just about the same amount of information as taking the genetic test: If you have two or more family members with the disease, you also have about a 35 percent chance of getting it. Or you can take one of the three GenoType Calculators (Basic, Intermediate, or Advanced) and find out whether you're in one of the GenoTypes that is most at risk for the disease. The latter two alternatives are just as effective as—and less expensive than—the lab test. The GenoType Calculator has a further advantage: Once you know your GenoType, you can follow up with your own individualized GenoType Diet to maximize your chances for avoiding the disease.

Third—and this is the most important point—*most disorders that we know about are not caused by a single gene; they're caused by the interaction among several genes.* We don't yet have lab tests to measure that type of interaction. But GenoTypes give us a clear picture of which of the six basic types of interaction you may be prone to, and which disorders you're at risk for as a result. So what the GenoType Calculators can do, which lab tests cannot, is give you a very good idea of many of your most important risk factors—and it can do so in a simple half-hour set of procedures that you can do at home, with no effort or expense.

## What Does the GenoType Calculator Do?

Whichever GenoType Calculator you choose to use (and I will walk you through that in a bit), it will ask you to measure or test certain aspects of your body—leg and torso length, the length of your index and ring fingers, and going a bit further, your blood type. Later on, you will also perform a few additional measurements and observations, such as the shape of your teeth, fingerprints, and a few others—that you will use to determine just how closely you fit your GenoType.

Remember, your GenoType is a survival strategy hammered out by your genes and your cells in response to your prenatal environment. We can read your prenatal environment in the length of your leg bones (indicating the presence of growth factors), the patterns on your fingerprints (indicating the levels of hormones), and other key signs. In and of themselves, most of these physical elements mean nothing: They themselves don't cause problems for you in later life, and they have nothing to do with your health and well-being. But they are symptoms of interactions that took place when you were in the womb—interactions that could predispose you to certain diseases and certain types of weight-loss issues.

Many of the physical characteristics measured by the calculators are like the golden coat of the Agouti mouse that you read about in Chapter 2. In itself, the golden coat meant nothing. But it happened to correlate with genes that did mean something—the genes for obesity that predisposed those mice to weight gain, diabetes, and many other obesity-related disorders. When the obesity genes were silenced in the womb by giving the mother a methyl-rich diet, the mice were born slender and healthy—and their coats were brown. There's no good reason why a brown coat would be associated with slimness while a golden one indicated obesity. But those genes happened to be linked, just as many of yours are. So by discovering telltale physical signs, we can read the interaction of genetic heritage with prenatal experience—and thereby discover your GenoType.

At first glance, the questions on this test may seem a bit daunting, particularly the ones that ask you about aspects of your body that you've never thought much about. You've probably never taken your own fingerprints or wondered whether your legs are longer than your torso. You've almost certainly never noticed whether the insides of your two front teeth are "shoveled" (scooped out), or whether you have an extra cusp (bump) on your first molar. So when you first read through the list of questions, you may find yourself feeling a bit overwhelmed.

Don't worry. The measurements are actually quite easy to do, and with the help of a friend, you can complete them within thirty minutes. If you want to get started on the test, skip right to Part II, where I'll talk you through every single measurement. You won't need anything more than a few simple items you probably have around the house, and you'll be surprised at how easy it all is.

But if you'd like to understand why I've included each of these measurements on the test—what fascinating information about you and your genetic potential these markers reveal—read on. You're about to find out more about those crucial nine months that produced the unique being that is you. Even better, you'll have the tools for discovering how to become the best YOU that you can be.

I will be spending the rest of this chapter discussing why these measurements and tests are critical to determining your GenoType. If you are more of a "doer," by all means feel free to move on to Chapter 4 and jump right into things. You don't really need to understand the science to determine your GenoType or to follow the GenoType Diet—though you may find yourself shaking your head in bafflement over how such seemingly minor information can reveal anything important about your body, your health, and the maintenance of your optimal weight. Don't worry: There's good, solid science behind every question. This chapter is for those readers who like to know the whys and wherefores behind something before they do it.

Whichever choice you make, I would like you to know how I designed these calculators. First, I wanted every question to be simple and easy to do at home. At the most elementary level (the Basic GenoType Calculator), the only things required are a ruler, a measuring tape, and some ink and paper for the strength-testing part. If you know your blood type, you can move on to the Intermediate GenoType Calculator and benefit from adding in these extra bits of information. To determine your GenoType at the level of the Advanced GenoType Calculator, you will need to know your ABO and Rh blood types and secretor status. (This calculator will be used primarily by readers of the Blood Type Diet series.)

Second, I wanted every question to have a clear, objective answer. You won't have to guess the answers to such vague questions as whether you "prefer to lead an active, vigorous lifestyle" or "are more comfortable spending a quiet evening at home." The answers to every single question should be clear and unmistakable, which means that you can't take this test "wrong" and you can't make any mistakes. Because GenoType information tends to cluster in recognizable patterns, there's a lot of built-in duplication—people with certain fingerprint patterns also tend to have certain body shapes, people with certain family histories tend to have certain finger-length ratios, and so on. So the questions you can answer will balance out the few that you may not be sure about.

Finally, every question is linked to an influential genetic process that can be affected by diet and lifestyle. Like the Agouti mouse's golden coat, every physical marker I ask you about indicates a biological process that you can modify by following your GenoType Diet. Every piece of information this test uncovers—from a tendency to put on weight to the increased risk of heart disease—will help you make informed, productive choices. You won't learn about any doomsday predictions you can't control. Instead, you'll be finding out about tendencies and potential threats that your GenoType Diet and Exercise Plan can help you address.

Of course, no diet can help you live forever or maintain eternal youth—at least, none that has yet to be discovered! But if you want the diet that maximizes your chances for long life, vitality, and optimal weight, you can find it in Part IV. And the route to that diet leads through the GenoType Calculators.

Now it's time to look at the actual calculators. Let me talk you through what you're going to find. The GenoType Calculator has two parts: *the calculator itself* and a section that *strength-tests your GenoType*.

Your first step is simply deciding which calculator (Basic, Intermediate, or Advanced) to use.

## Choose Your GenoType Calculator

The first step in learning your GenoType is to decide which calculator is best for you. The GenoType Diet can grow with you. Perhaps today you will use a simple calculator to get up and going right away. Or you may be one of the readers of my series of Blood Type Diet books and will be able to skip the elementary level and begin your journey with a slightly more precise diagnostic tool. No matter what level you begin the Geno-Type Diet program at, you can always return when you have more information and use a more sophisticated calculator.

Each calculator has a certain level of information that it requires. The Basic GenoType Calculator requires two simple body measurements, while the Advanced GenoType Calculator, in addition to these two measurements, requires your blood type and secretor status.

So let's get started with a quick guide on how to decide which Geno-Type Calculator is right for you.

- If you do not know your ABO and Rh blood types, you will be completing the Basic GenoType Calculator.
- If you know your ABO blood type, you will be completing the Intermediate GenoType Calculator.
- If you know your ABO and Rh blood types and your secretor status, you will be using the Advanced GenoType Calculator.

Regardless of which calculator you choose, everyone will do the two basic body measurements and also complete a series of questions, measurements, and observations to allow you to determine just how strongly you fit the total picture of your GenoType. I call this "strength-testing" your GenoType, and it gives us a good idea of just how to structure your GenoType Diet and Supplement protocol.

## The Basic GenoType Calculator

At its most elemental level, you can find out your GenoType by:

- Measuring the length of your torso and legs
- Measuring the length of your index finger and ring finger on both hands
- Completing the requirements for strength-testing your GenoType
- Running the basic GenoType Calculator to determine the four top GenoType candidates for you
- Strength-testing both candidates to determine the GenoType that most closely matches you

I will teach you the proper methods to perform these measurements in the next chapter. Right now, I thought you might enjoy learning a little bit of the science behind their inclusion in the GenoType Calculators.

## Measurement #1
### Which is longer, your torso or your legs?

I'll talk you through the actual measuring process in Part II. For now, let me say that leg and torso ratios reflect the levels of the growth factor hormones, in particular, insulin-like growth factors 1 and 2 (IGF-1, IGF-2), that you encountered in the womb and early childhood. Insulin-like growth factors are especially significant because they're strongly associated with growth and height.

Now, here's where it gets interesting. Leg length and height in general correlate to the risks for various diseases. For example, being short and having short legs appears to increase your risk for being overweight and for developing Type 2 diabetes in middle age. Short leg length in general is correlated with your risk for coronary heart disease.

Being tall, on the other hand, seems to increase your risk for cancer, especially such hormonally dependent cancers as breast and prostate, which seem to be associated with high levels of IGF-1. It may be that tall people hit puberty earlier and so spend more time being exposed to adult concentrations of sex hormones, which may lead to prostate cancer in men and breast cancer in women.

Legs that are equal to or shorter than your torso are indicators for Teacher and Explorer GenoTypes, who tend to have a very low center of gravity. Since these folks had to till the soil before we developed the agricultural technology to ease their task, maybe they were built for shoveling, lifting, and other types of heavy labor.

When a patient of mine heard that short people were at risk for heart disease and diabetes while tall folks were at risk for cancer, she asked me somewhat plaintively if there wasn't an ideal height that would present fewer health risks. I replied by quoting the words of my favorite oncology professor, who may have been quoting Jim Morrison: "Nobody gets out of here alive." We're all going to die someday—the best diet in the world can't prevent that. However, the right diet for your GenoType *can* extend your life and vitality, minimizing your risk for the diseases that tend to attack your type.

From that point of view, it's good to know that these ratios do help predict GenoType, since that will steer you to the healthiest possible diet.

When we get to the actual measurements in Part II, I'll also teach you how to tell if your lower leg is longer than your upper. You don't need it for the Basic GenoType Calculator, but you will need it to answer some of the strength-testing questions, so let's get it done while your friend is over and you have everything in place. Longer upper legs are a hallmark of the Gatherer and Nomad GenoTypes, while longer lower legs are a keynote for the Warrior and Hunter GenoTypes.

For those of you wondering about what happens if the lengths are equal, you can relax: Ties always go to the torso and lower leg.

## Measurement #2
### Which is longer, your ring finger or your index finger?

This seemingly trivial detail—index-to-ring-finger ratio, or D2:D4 in scientific terms (the "D" stands for digit)—is actually an excellent marker for sex hormone exposure. A longer ring finger means you encountered more androgens in the womb (androgens are a testosterone precursor); a longer index finger means you faced higher levels of estrogen.

Comparing the results from hand to hand also gives some indication of symmetry, the degree of "sameness" from one side of the body to the other. Asymmetry is a powerful indicator of developmental stress while in utero. D2 and D4 ratios have also been linked to the role of a critical group of genes know as Hox genes. Found in almost every species, Hox genes control how the body creates its "segments," such as the head, neck, chest, and abdomen, so looking at finger length can tell you a lot about yourself.

Longer ring fingers on both hands are a keynote for the Hunter GenoType, though people in other GenoTypes have this pattern as well, including the Explorer and Nomad GenoTypes. More male hormones tend to contribute to a more "andric" (male) shape: longer, leaner, more muscular. However, with the Explorer, these patterns are usually *asymmetrical to gender*—in other words, you will see longer ring fingers in Explorers, usually if they are female. GenoType 6, the Nomad, is usually *symmetrical to* gender, meaning that you can find longer ring fingers in male Nomads and longer index fingers in female Nomads.

Having longer index fingers on both hands is a hallmark of the Gatherer GenoType, although it is also seen in female Nomads and occasionally in Teachers. That's probably why Gatherer GenoTypes are more "gynic" (female-shaped) and rounded.

A different result on each hand (that is, having a longer ring finger on one hand and a longer index finger on the other) is a hallmark of the GenoType 3 Teacher, although it is also seen in GenoType 2, the Gatherer.

So if the Basic GenoType Calculator is your calculator of choice, you can just move on to the section that describes the tests you will use to strength-test your GenoType. If you are planning on using the Intermediate or Advanced GenoType Calculator, read on.

## The Intermediate GenoType Calculator

At this level, you can find out your GenoType by:

- Measuring the length of your torso and legs
- Measuring the length of your index finger and ring finger on both hands
- Completing the requirements for strength-testing your GenoType
- Testing yourself (or finding the information) for your ABO blood type
- Running the Intermediate GenoType Calculator to determine your GenoType
- Strength-testing both candidates to determine the GenoType that most closely matches you

Now we're up to the first of the "classic" genes—the single genes that tell us so much about who we are and what we need. These genes are classic because they've been part of the genetic landscape since the moment of their discovery and, despite all the innovations in recent years, still play a critical effect in our understanding the effects of genes and health.

## Test #1
### What is your ABO blood type?

The Intermediate GenoType Calculator is where you can correlate your GenoType with your blood type, especially if you've already learned about your blood type from my earlier book, *Eat Right for Your Type*. People in blood group O are most likely to be Hunter or Gatherer GenoTypes, and a few are Explorers. People in blood group A are most likely to be Teacher or Warrior GenoTypes, though some are Explorers. People with blood group B are most likely to be Nomad GenoTypes, though a few are either Gatherer or Explorer Genotypes. And people in blood group AB are most likely to be Nomads or Warriors, though some are Explorers and a few are Teacher GenoTypes.

As you can see, the Explorers GenoType is really the catchall group, the Swiss Army knife of the GenoTypes: They are the only GenoType that includes members of all four blood groups. On the other hand, if you are Rh-negative, you are most likely to be an Explorer as well.

If you've read *Eat Right for Your Type*, you already know that various disorders correlate more strongly with some blood types than others. Blood type O is more vulnerable to cholera and the plague; type A to smallpox; type B to many types of influenza; and type AB to malaria. Blood types are also correlated with levels of stomach acid, responses to lectins and glutens (found in wheat and other grains), and the types of bacteria that are at home in your digestive tracts. As a result, different types of diets suit different blood types, with type O needing a more carnivorous diet, type A tending toward a vegetarian approach, and type B falling somewhere in between. If you are intrigued about the role blood type may play in your chances of developing certain diseases, I recommend that you read my books *The Complete Blood Type Encyclopedia* and *Live Right 4 Your Type*.

As we saw in my previous books on blood type, your blood type doesn't cause your immune reactions or your digestive issues. Rather, like the golden coat of the Agouti mouse, it simply correlates with them. Rather than being an epigenetic influence itself, blood type influences the environment in which those epigenetic decisions are made. For example, when you are but a week old, your blood type is busy helping to determine when your budding circulatory system should begin placing your arteries, much like a surveyor always moving ahead of the work crew, telling them where to put the highway.

As we've seen, the six GenoTypes represent six different survival strategies. Some of the signs of these strategies have obvious purposes, such as whether your system is reactive (geared to protect against infectious diseases), thrifty (geared to hold on to calories), or altruistic (geared to tolerate a wide variety of environments). Some have obvious sources, such as the presence of growth factors that stretch your bones or the presence of sex hormones that affect the length of various fingers.

Blood type is in that last category. It doesn't affect your health per se. Nobody dies just because they have a certain blood type. But as a sign of what's going on in your system, it correlates with so many other important pieces of information that it's one of the most significant physical markers we have.

So if the Intermediate GenoType Calculator is your calculator of choice, you can move on to the section that describes the tests you will use to strength-test your GenoType. If you are planning on using the Advanced GenoType Calculator, read on.

## The Advanced GenoType Calculator

At this level, you can find out your GenoType by:

- Measuring the length of your torso and legs
- Measuring the length of your index finger and ring finger on both hands
- Completing the requirements for strength-testing your GenoType
- Testing yourself (or finding the information) for your ABO and Rh blood types
- Testing yourself for your secretor status

- Running the Advanced GenoType Calculator to determine your GenoType
- Strength-testing candidates to determine the GenoType that most closely matches you

At the Advanced level, you will be factoring in your secretor status, in addition to the measurements and tests found in the Basic and Intermediate Calculators.

## Test #2
### What is your Rh blood type?

Although clinically important, the Rhesus (or Rh) blood group system is not as widely distributed in the body fluids and tissues as the ABO blood groups. However, it is an important genetic and anthropological marker. In the GenoType Diet, it is used as part of the Strength Testing for the Explorer GenoType as well as in the Advanced GenoType Calculator.

## Test #3
### What is your secretor status?

This test is required for the Advanced GenoType Calculator.

If you would like to use the Advanced GenoType Calculator and need to determine your secretor status, please refer to the Resources Section in the back of the book for information on how to purchase a Secretor Submission Kit. Testing for secretor status is not as easy as testing yourself for your ABO or Rh blood type; you have to send a saliva sample to a lab for analysis. However, the test is not expensive compared to most genetic testing, and the information your secretor status can provide to you can be truly life-changing and life-saving. Please note that once you've sent in your saliva sample, it takes approximately three weeks to get your results.

Secretor status is closely linked to your ABO blood type, since the secretor gene controls whether or not you "secrete" your ABO blood-type antigen in your body secretions. Whereas all of us could be blood-typed from a slide of blood, about 85 percent of us secrete our blood type in a free form that is pumped out into our body secretions, sweat, semen, mucus, saliva, and so forth. These people are called *secretors,* and if we wanted to, we could probably blood-type them from a saliva or semen sample as well. About 15 percent of the world is missing the secretor allele, and not surprisingly, they are called *non-secretors.*

Secretor status has important effects on our metabolism and immune function. Non-secretors, for example, have lower levels of fat-busting enzymes, more environmental sensitivities, weaker defenses against yeast and parasites, and a tendency for inflammatory problems such as arthritis.

## Strength-Testing Your GenoType

In this section, I'll describe some of the measurements you may need to complete in Chapter 5 to see just how good a fit you have with the Geno-Type that you've calculated for yourself. In addition to providing some fascinating insight into what makes you uniquely you, strength-testing helps you determine the degree of compliance that you should devote to your GenoType program. Finally, if you are doing the Basic GenoType Calculator, the information in this section allows you to make the final selection from the four candidate GenoTypes that the Basic Calculator suggested.

I will teach you the proper methods to perform these measurements and observations in the next chapter. Right now, I thought you might enjoy learning a little bit of the science behind why they can help determine your GenoType.

### The Five Basic Questions

Unlike a few other diets, the GenoType Diet system does not use "questionnaire-type" questions to arrive at your GenoType. These types of questions are notoriously subjective, meaning they will often mean different things to different people. The Five Basic Questions used to strength-test your GenoType are simple yes-no questions that leave very little room for interpretation—they either apply to you or they don't.

## Question #1

- **Are you sensitive to caffeine? Would a cup of coffee in the evening keep you awake at night? Or are you insensitive?**

Sensitivity to caffeine is a classic sign of the Explorer GenoType. That's because many Explorers have a gene that makes them what we call a slow acetylator. You don't need to remember the technical term. But you might like knowing that *acetylation* is the chemical process your liver uses to detoxify any foreign element that makes its way into your body. Drugs, alcohol, and even prescription medications are all read by your liver as toxins that must be cleared from your system. People with "fast acetylator" genes perform this detoxification swiftly and efficiently. They're the ones who can really hold their liquor, who aren't so sensitive to medication, who generally don't get food poisoning or have intense reactions to cigarette smoke. The Warrior GenoType is typically a fast acetylator. Fast acetylators have their own set of problems—for example, they don't seem to break down the carcinogens in cooked meats too well, which may increase their risk for colon cancer.

Slow acetylators, by contrast, don't do well at cleansing their bodies of these foreign substances. One drink lays them out; the medication dosage that most people take makes them dizzy, nauseated, overstimulated, or sleepy. While a few slow acetylators can be found across every GenoType, they tend to predominate among Explorers.

## Questions #2, 3, 4, 5

- **Among you, your parents, grandparents, and siblings, have there been two or more instances of clinical depression or cognitive dysfunction such as Alzheimer's disease?**
- **Among you, your parents, grandparents, and siblings, have there been two or more instances of heart disease, stroke, or diabetes?**
- **Among you, your parents, grandparents, and siblings, have there been two or more instances of cancer?**
- **Among you, your parents, grandparents, and siblings, have there been two or more instances of autoimmune disease (lupus, rheumatoid arthritis, multiple sclerosis)?**

Again, these diseases may strike anyone in any GenoType, but they tend to correlate to particular GenoTypes. Some of this is statistical: I can show you evidence of how the disease patterns fall. Some of it is biological: I can explain why some characteristics of each GenoType might predispose people to certain disorders.

For example, the Hunter GenoType tends to have extremely reactive immune systems that produce inflammation at the drop of a hat. Inflammation (the increased production of white blood cells, usually accompanied by heat, pain, swelling, and redness) is the body's response to any perceived toxin or threat. When the body is facing an actual threat—the virus for the common cold, for example, or the potential infection from a cut or wound—inflammation is extremely useful in overcoming the invader. When the body is overreacting—to the dust particle that sets off an asthma attack or the peanut that sends you into anaphylactic shock—the Hunter's reactivity is counterproductive. A family history of autoimmune disease can indicate Hunters in your family, making it more likely that you've inherited their reactive tendencies.

On the other hand, Teacher GenoTypes are just the opposite: Their altruistic immune systems tend to welcome everything in their environment—even cells that may be working their way toward cancer or bacterial invaders, both of which they'd do better to repel. So a family history of cancer might be a sign of the Teacher GenoType.

The GenoTypes most vulnerable to heart disease, stroke, and arterial disorders are the so-called thrifty GenoTypes: Gatherers and Warriors. Warriors appear to have blood that is naturally more viscous (thick) and arteries that by nature may be more prone to degenerative changes. Gatherers often have family lineages loaded with adult-onset diabetes. The Achilles' heel of GenoType Nomads is the connection between their nervous system and their immune system. Like Teachers, Nomads are altruistic, but they're far better than Teachers at identifying certain potential threats. This GenoType seems to have difficulty when it comes to communication among the various systems of the body. Their problems are less with the immune system or the nervous system, and more with the communication between them. As a result, they present a fairly idiosyncratic set of defenses: Although they seem impervious to most diseases, they have a few very specific vulnerabilities that wily predators can take advantage of.

## Fingerprints: A Developmental Road Map

Remember, epigenetics is the science of what happens at your town meeting: the interaction between the genes, your environment, and your diet. One important chapter in your town meeting's history takes place in the womb. So you can think of your fingerprints as the minutes of that meeting, your record of the deals that were worked out among your genetic heritage, your prenatal diet, and the environmental signals you received in the womb.

Fingerprints are one of those signs that don't mean anything by themselves—they don't "give" you cancer or put you at risk for heart disease. But since they correlate so strongly to events in your epigenetic history, they can be used to determine potential threats to your health, including rheumatoid arthritis, heart disease, diabetes, and cancer. They can also be used to identify your GenoType.

The technical name for fingerprint analysis is *dermatoglyphics*, but unless you're looking for work in a crime lab, you don't need to remember that term. However, you may find it helpful to envision how your fingerprints were formed, since that gives you a better picture of your epigenetic history.

When you're just a little fetus only six and a half weeks old, you don't yet have fingertips; instead, you have volar pads, which continue to grow until the end of your first trimester. Then they begin to shrink, and the bones that become your fingers are soon covered in flesh marked by unique raised areas that turn into fully formed fingerprints by your twenty-first week.

As a result, every major event between Week 6 and Week 21 of your fetal life leaves its mark in your unique pattern of loops and whorls. In fact, that's why identical twins don't have identical fingerprints. Even though their genetic heritage is exactly the same, their prenatal experience differs: One twin gets more food, the other may have been more affected by Mom's stress hormones, or by that time she had slipped and fallen down. Epigenetics is the reason that identical twins aren't actually identical.

You may be wondering whether your fingerprints could actually be correlated to specific incidents that took place while your mother was pregnant. They can't, because they're recording not a set of discrete events but rather a whole process. Think again of that town meeting. Suppose a topic is introduced: "We need to respond to the lack of nutrients," or "We need to respond to this flood of stress hormones." Your genes, environment, and diet begin a debate: Should we respond by becoming more reactive? More thrifty? More altruistic? Maybe we should learn to detoxify caffeine more quickly. Maybe we're stuck with detoxifying it more slowly.

Meanwhile, diet and environment keep bringing new information into the debate, affecting it accordingly. Finally, a conclusion is reached. That conclusion is recorded in your fingerprint patterns, the resolution that was finally taken at the meeting. By reading the resolution, you can infer what diet and environmental factors came into play and what epigenetic decisions were made that steered you toward your current Geno-Type. You can find out which disorders you're more prone to and which biological strengths you developed. And then you can use this information to figure out which diet will most help you maximize your strengths and minimize your weaknesses.

Because they're such a good record of your prenatal life, fingerprints are an important clue both to your GenoType and to the disorders that correlate with it. In fact, there are thousands of studies correlating fingerprint patterns with potential health risks. For example, numerous studies have found that patients with fingerprints containing eight or more instances of a pattern called "ulnar loops" have a tendency to Alzheimer's and cognitive disease in later life. Eight or more ulnar loops are also one of the keynotes for GenoType Nomad—the very group that has a strong tendency to Alzheimer's and cognitive disease. (Of course, I'll show you how to take your fingerprints and how to identify the significant patterns in Part II, where you'll also find detailed diagrams against which you can compare your own fingerprints.)

Likewise, six or more of the fingerprint patterns known as whorls have been statistically correlated with a higher risk of breast cancer. These whorls are also a keynote for GenoType Teacher, who may face a higher-than-average risk of cancer. In fact, six or more whorls have about the same diagnostic validity as a positive mammogram and a positive breast biopsy.

Finally, GenoType Warriors, who tend to have more than three arch-type fingerprint patterns, should be aware that there seems to be a correlation with sluggish bowels and problems with elimination.

As you can see, knowing about these potential problems makes following your GenoType Diet and Exercise Plan even more important, because they give you the best possible shot at avoiding these potential dangers, much as a traffic sign warns you to slow down or pump your brakes to avoid a curvy or slippery stretch of road. Eating the right diet for your GenoType can help you beat these epigenetic odds, ensuring your greatest chance for a long and healthy life.

## Symmetries: What Your Fingers Can Reveal

Some GenoTypes are more or less symmetrical than others. As I said in our little talk about finger length, left-to-right differences in the body are often a sign of developmental stress, whereas symmetry tends to correlate with fitness. Generally, the more your left side resembles your right, the fitter you are and the less stress you experienced in the womb. The more asymmetrical you are, the rougher the time you probably had as a fetus. That's because the left and right sides of your body tend to develop separately: Cells working on one side of the body don't actually know what their counterparts on the other side are doing. All things being equal, the two sides will follow the same logic and develop identically. But when stress interferes with the fetus's development, it's likely to do so in uneven ways—and asymmetry is the result. So a simple formula is: The more stress, the less symmetry.

As we've already seen, one way of checking for symmetry is to check the finger lengths and compare the right and left hands. Another method is to "book-match" the patterns by comparing the pattern on each finger with that finger on the opposite hand. For example, does the pattern on the index finger of the right hand match the pattern on the index finger of the left?

The more symmetrical GenoTypes, such as GenoType Hunter and GenoType Nomad, will often have four or five out of five fingerprints matching from hand to hand. On the other hand, the more asymmetrical GenoTypes, such as GenoTypes Gatherer and Explorer, will tend to have three or more fingers that don't match. An especially unique characteristic of the Explorer is that the index fingers typically don't match; they might have a whorl on the index finger of the left hand and a loop on the index finger of the right.

Handedness is also a reflection of the fetal environment. Again, this isn't terribly significant, but Explorer GenoTypes tend to be more left-handed or ambidextrous than most other GenoTypes. Left-handedness and ambidexterity are signs of fetal stress, possibly caused by hormonal shifts during pregnancy.

## Biometrics: The Physical Imprint of Your Time in the Womb

*Biometrics* is literally "the measure of living things": It's a way of measuring your morphology and other key elements of your physical self.

There are two major categories that help define this section. One includes the elements that indicate male, or andric, shapes: exposure to testosterone and a high number of prenatal hormones known as growth factors. The other includes elements that indicate female, or gynic, shapes: exposure to estrogen and a lower number of growth factors.

The classic andric GenoType is the Hunter; the classic gynic profile is the Gatherer. Of course, we've all known tall, rangy women and more rounded men; they've simply had higher-than-usual prenatal exposure to the "opposite-sex" hormone.

We're also interested in some specific issues that you've probably never thought about before: the ratios of your lower and upper legs and the ratios of your legs to your torso. Both figures indicate the presence or absence of prenatal growth factors, which in turn are linked to various types of strengths and weaknesses.

It was shown in the 1940s that the space between the upper legs right above the knees is a good indicator of andric versus gynic tendencies. Meanwhile, I can confirm that, as you've probably guessed, a small opening is more typical of women and is in fact a keynote for the Gatherer GenoType, whose high levels of prenatal estrogen and thriftiness contribute to their rounded bodies. (If you're a guy with a Gatherer Geno-Type profile, and there are many, don't worry—you're no less manly. You just have a different set of health concerns to watch out for and you'll need to adjust your diet accordingly, as you'll read in Part IV.)

On the other hand, larger openings are more characteristic of Geno-Type 1 Hunters and, to some extent, GenoType 5 Warriors. Both types tend to have longer bones and, especially, longer legs, because of the increased presence of growth factors in their prenatal environment. You'll find a diagram in Part II that will make it easier for you to tell whether the opening between your upper legs is large or small.

Another simple but useful biometric marker is the visible presence of tendons and sinews under the skin in your limbs—particularly at the wrists, a keynote for the Teacher GenoType, who tends to have long, sinewy limbs. Just think of the long, wiry arms of Abraham Lincoln, a classic Teacher type. Gatherers, on the other hand, tend to look "padded" even in places where there is not much fat.

Perhaps you have heard of the common body type descriptions—the rounded *endomorph*, the lanky, lean *ectomorph*, and the muscular *mesomorph*. When you think about which of the three body types you have, you may know immediately, or you may need to look at the questions and illustrations in Part II, which will help you determine this attribute and decipher the technical terms in parentheses. Again, if you're currently overweight—or if you think you are—you may not correctly identify your body type. I'm always struck by the number of women who think they're heavier than they are, so this may be one of those questions where a friend can help you come up with a more accurate answer.

Generally, a *truly* rounded body type is a keynote for GenoType Gatherer, though GenoType Warrior, with their thrifty genes, also tend in this direction, especially as they age. Explorer GenoTypes, and some Nomad GenoTypes, are often muscular, whereas GenoTypes Hunter and Teacher tend toward the compact or lanky. Again, these correlations probably reflect both family genetics and prenatal exposure to sex hormones and growth factors.

In and of themselves, body shapes don't necessarily correlate to health issues, though they often do correspond to metabolism, which in turn reflects the ease with which you gain, lose, or retain weight. Generally, if you're an ectomorph—a thin, lanky body with small bones—your metabolism burns higher and you'll find it fairly easy to maintain a slender size. Muscular, broad-shouldered mesomorphs have a medium-to-high metabolism, burning off the calories quickly, while big-boned, rounded endomorphs tend to have slow metabolisms and, often, thrifty genes that hang on to calories and hold on to fat. Following the correct GenoType Diet can help you get your metabolism running at the speed that is right for you, making it far easier to achieve your optimal weight.

## Dentition: Getting Your Teeth into Things

Teeth can tell a lot about a person. We will be interested in two dental landmarks: whether you have spade-like front incisor teeth called *incisor shoveling,* and whether you have an extra cusp (bump) on you first molar, called *Carabelli's cusp.*

A Japanese researcher has discovered that tooth shoveling is a fairly good sign that you've got an ancestral history of either herding or hunting, leading us to speculate that perhaps those shoveled-out front teeth are better equipped at tearing meat. Accordingly, it's a keynote for GenoType Hunter and a lesser indicator for GenoType Explorer.

The extra molar cusp seems to help you grind food better, perhaps reflecting the more early agrarian diets of the Teacher and Warrior GenoTypes and the hardscrabble subsistence of the early Gatherer GenoTypes.

You'll find illustrations in Part II that will help you determine whether your teeth are "shoveled" and if you have the extra molar cusp, though you may need a friend to have a look in your mouth with a pocket flashlight.

## Waist Not, Want Not

In Chapter 5, you will learn how to determine your waist-to-hip ratio, which is just comparing the measurement of your abdomen at the waist to the measurement at the hips. These ratios are a better indication of biological health than the more commonly used body mass index (BMI) (based on the ratio of height to weight). I don't like to use the BMI, because it doesn't distinguish between men and women. I also find that the waist-to-hip ratio is a better indicator of the risk of heart attack. When obesity is redefined based not on the BMI but on the waist-to-hip ratio, the proportion of people at risk for heart disease increases threefold. Plus, it's far easier to get an accurate waist-to-hip ratio—just two measurements, one calculation, and you're done. (See Part II.)

For women, an ideal waist-to-hip ratio indicates an optimal exposure to estrogen and therefore correlates to health and fertility. Women with an ideal waist-to-hip ratio tend to be less susceptible to cardiovascular disease, diabetes, and ovarian cancer.

Men with an ideal waist-to-hip ratio are also healthier and more fertile, having received the optimal exposure to androgenic hormones such as testosterone. As a result, they're less susceptible to prostate and testicular cancer.

An ideal waist-to-hip ratio—an hourglass shape in women and a square-hipped shape in men—is a keynote for GenoTypes Hunter and Teacher and some GenoType Nomads. It indicates that throughout your life, your tissue is less sensitive to estrogen. That lack of sensitivity tends to increase your risk of heart disease and osteoporosis while reducing your risk of reproductive cancers.

A high waist-to-hip ratio—a more boxy shape, with a less defined waist—is a keynote for the rounded, thrifty-gene Gatherer and Warrior GenoTypes.

## Head Shape and Jaw Angle

Jaws come in three basic varieties: wide-angled, giving the face an almost almond shape; narrow, giving what some call a "lantern jaw"; and everything else that's in between.

The shape of the jaw is determined by what anatomists call the *gonial angle.* Wide gonial angles give an almond-shaped profile and are a keynote for GenoType Gatherer, but they are also seen with the Warrior and Nomad GenoTypes. Narrow gonial angles give a squarish jaw that tends to be more common in the Hunter, Teacher, and Explorer Geno-Types. Jaw angles reflect prenatal growth factors, and they seem to be the result of prenatal regulators of adhesive tissue—literally, tissue that helps things stick together. Statistically, an almond-shaped jaw seems to correlate to such disorders as toxemia during pregnancy, pernicious anemia, ulcers, and migraines in women, and to migraines in men.

Square- or lantern-shaped jaws, on the other hand, may reflect more slippery tissue, and that may be why they're more correlated with breast and uterine cancer: The cancer cells have an easier time detaching from slippery tissue and then spreading through the body. Square jaws also correlate to gallbladder problems. You'll find illustrations in Part II that will help you determine whether you have a wide or narrow gonial angle.

Head shape is another developmental feature that correlates with GenoType. Head shape can be squarish, elongated, or a blend of both. The elongated head shape is a keynote of the Warrior GenoType, whereas squarer heads are found in most Explorers and many Nomads. Human head shapes are gradually becoming more elongated. This elongation seems to correlate with the increase in height seen in modern humans. Up until the Middle Ages, on the other hand, heads had been getting broader, so there must be some correlation with changes in diet and hygiene.

You'll find diagrams in Part II that will help you determine your head shape.

# What Is Your Taster Status?

Being a *taster* implies that you have a gene that enables you to taste a compound called phenylthiocarbamide (PTC), which occurs in cruciferous vegetables, such as broccoli, cabbage, and cauliflower. Because PTC is not an altogether safe substance, a similar compound known as propylthiouracil (PROP) is what's used in laboratory tests.

PTC and PROP taste bitter—if you can taste them. Some people, known as "non-tasters," actually can't, and indeed, not being able to taste this substance is a hallmark of the "thrifty" GenoTypes, the Gatherers and Warriors. Some people are completely repelled by the taste, including the Hunter and Explorer GenoTypes and a few Teacher GenoTypes. These individuals are called "super-tasters." Still others can taste the substance but don't have an especially strong reaction to it. They are called "tasters." That group includes many Teacher and Nomad GenoTypes and some Gatherers as well.

Altogether, about 70 percent of all humans can taste this substance, but that range varies—from 58 percent of aboriginal peoples to 98 percent of Native Americans. Possibly the genes influence food selection by making some foods more palatable than others. Studies indicate that tasters—especially super-tasters—have a negative reaction to foods with strong tastes, including sweets, bitter foods, fatty foods, alcohol, coffee, and tea. Super-tasters may also develop more rapidly at puberty than non-tasters, probably as some sort of evolutionary adaptation to get to child-rearing and procreation ages as soon as possible, which may explain why super-tasters are so often found in Hunters and Explorers. As always, there are trade-offs. Super-tasters may avoid cruciferous vegetables—and thereby miss out on their cancer-inhibiting effects. On the other hand, non-tasters may suffer from low thyroid function, perhaps because the chemicals involved in the tasting mechanism are thyroid inhibitors. Perhaps because *their* thyroid glands function well, tasters tend to have a higher ratio of muscle to fat than non-tasters, as well as fewer weight problems. Then again, some tasters and even more super-tasters suffer from overactive thyroid. As always, the goal is balance, which your Geno-Type Diet can help you achieve.

How do you know whether you're a taster? If you order a GenoType test kit (see Resources), you'll get a tasting strip and a control.

## Preparing to Use
## Your GenoType Calculator

Now that you understand the science behind the GenoType Calculators, you're ready to do the actual work. So turn to Part II and prepare to learn a lot more about yourself. All it takes is a few household items, a half hour of your time, and the willing participation of a friend. What you'll gain in return is some invaluable information about who you are—and who you might become.

# PART II

# Understanding Your Body's Clues

## Why You Don't Need a Genetics Lab

# CHAPTER FOUR

# Using the GenoType Calculators

The **GenoType Diet** grows with you. Perhaps you know your blood type, or maybe you don't—yet. As you add new information about yourself, the picture of your GenoType will come more and more into focus. You can start your GenoType Diet journey with some simple body measurements and thirty minutes of your time.

But I suspect that most people will want to increase the accuracy of the GenoType outcome by factoring in more information. Many of you already have enough information about yourselves, such as your blood type, to skip the most basic calculator altogether.

Regardless of whether you use the basic calculator or even the most advanced, you can always check the strength of your results by completing a GenoType-specific Strength Meter in Chapter 6.

## The Three GenoType Calculators

- The Basic GenoType Calculator is the simplest, quickest, and easiest to do. It will get you roughly into the right ballpark. The Basic Calculator requires two measurements.
- The Intermediate GenoType Calculator utilizes the two measurements in the Basic GenoType Calculator as well as your ABO blood type. **The majority of the readers of this book will be using this calculator to determine their GenoType.**
- The Advanced GenoType Calculator is for those who already know their ABO and Rh blood types and secretor status. I'll explain in further detail what these are later on in this chapter.

If you are new to all this, you can start with the Basic GenoType Calculator and get going right away. In time you may want to add more information about yourself and move up to the Intermediate Calculator. Not only is this easy to do, but the tests that you will need to run on yourself are inexpensive and commercially available (see Resources for more information).

## Strength-Testing Your GenoType

After you have decided which GenoType calculator you want to use and done the measurements and/or collected the data needed, you can move on to the next chapter and "strength-test" your GenoType. Strength-testing your GenoType allows you to see how closely you personally fit the description of your GenoType. This section involves taking your fingerprints, answering some questions about yourself, and making a few additional simple body measurements. When you are done with this section, you can then move on to the next chapter and find out your GenoType.

# Which GenoType Calculator
# Is Right for You?

The first step in taking your GenoType test is to decide which of the three calculators—Basic, Intermediate, or Advanced—is best for you:

- If you are new to this and don't know your ABO (A, B, O, AB) and Rh (positive or negative) blood types, you can start with the Basic GenoType Calculator.
- If you know your ABO blood type, you will take the Intermediate GenoType Calculator.
- If you already know your ABO and Rh blood type and your secretor status, you can take the Advanced GenoType Calculator.
- No matter which GenoType Calculator you use, you will then move on to the next chapter and perform the tests and measurements necessary to strength-test your GenoType.

# Getting Started with the
# Basic GenoType Calculator

The Basic GenoType Calculator requires only two body measurements. After completing this section, you can move on to the next chapter.

*To perform the tests necessary for the Basic GenoType Calculator, you will need:*
- Fifteen minutes, with a friend present for most of that, to help you measure
- A calculator to do some very simple math
- A firm measuring tape, the kind you can use to measure your height
- A small ruler that you can use to measure your finger length in millimeters
- A standard kitchen chair with backrest

## Measurement 1:
## Leg and Torso Length

As we saw in Chapter 3, leg and torso ratios reflect the levels of the growth factor hormones that you encountered in the womb. These ratios correlate strongly to various types of diseases—and also to GenoTypes. That's actually good news, because once you've identified your GenoType, you can use your GenoType Diet to help ward off diseases to which you may be particularly susceptible.

You will be answering two basic questions: First, is your torso (the central part of your body, plus your neck and head) longer than your legs, or are your legs longer than your torso? The second question, and its result, will be used in the next chapter. Is your upper leg (the bone from the knee to the hip) longer than your lower leg (the bone from your knee to your ankle), or is your lower leg longer than your upper?

### How to do it

If you are uncertain of the location of any of these measurements, have a look at the drawing on page 59.

**1. Determine your standing height.** Stand straight, in your stocking feet, with a book on your head to make sure your head is level. Have your friend measure from the bottom of the book straight down the floor, ideally using a rigid tape measure. You can either hold the edge of the tape measure against the bottom of the book, or your friend can step on the tab of the tape measure and pull it up to the bottom of the book. The tape measure's case will be part of the measurement, so look on the case to see how many inches to add. Record your standing height on a piece of paper.

**2. Determine your sitting height.** Sit in a standard kitchen chair, again with the book upon your head, and have your friend measure from the bottom of the book down to the ground. Record your sitting height.

**3. Determine the height of a standard kitchen chair.** Using your tape measure, measure from the top of the chair seat straight down to the ground. Record the chair height.

**4. Subtract the chair height from your sitting height.** That's the length of your torso. Record it.

**5. Calculate your total leg length.** Subtract your torso length from your standing height. Record your total leg length.

**6. Measure your lower leg.** As you stand straight, in stocking feet, have your friend measure up the side of your leg, starting from the place where your ankle sticks out the farthest and ending at the bump on the outer side of your leg, just below your kneecap. If you're carrying excess weight, you may have trouble finding the bump; if so, just flex your knee until your friend can feel it. Then have her keep her hand there and measure down. Record your lower-leg length.

**7. Calculate your upper-leg length.** While sitting in a chair, place your fingertips on the kneecap of either leg. Slide your fingers upward until they are off the kneecap and your fingertips slide into the groove above the kneecap. Make a mark at this spot with a washable marker. Now look for the long skin crease between your hip and your leg. From this point, meas-

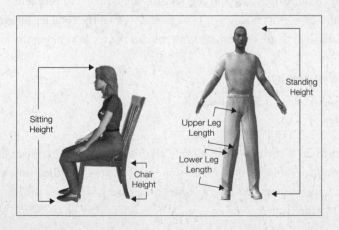

ure in a straight line to the mark you made on your knee. This is your upper-leg length.

### Record Your Leg and Torso Lengths

| 1. Standing Height: | | | | | |
|---|---|---|---|---|---|
| 2. Sitting Height: | *minus* | 3. Chair Height: | *equals* | 4. Torso Length: | |
| 1. Standing Height: | *minus* | 4. Torso Length: | *equals* | 5. Total Leg Length: | |
| 6. Lower-Leg Length: | | | | | |
| 7. Upper-Leg Length: | | | | | |

OK, let's organize these measurements into something we can save for later use.

Which is longer?

| 4. Torso Length: | 5. Total Leg Length: |
|---|---|
| ☐ Longer | ☐ Longer |

What to do if the torso and leg lengths are equal? No problem: "Ties" (equal measurements) always go to the torso, so if the total leg length and the torso length are the same length, record the torso as being the longer.

Which is longer?

| 6. Lower-Leg Length: | 7. Upper-Leg Length: |
|---|---|
| ☐ Longer | ☐ Longer |

What to do if the upper-leg length and the lower-leg length are equal? No problem: "Ties" (equal measurements) always go to the lower leg, so if the upper-leg and the lower-leg lengths are equal, record the lower leg as being the longer.

## Measurement 2:
## Index- and Ring-Finger Lengths

Another hormonal issue concerns the relationship between your ring finger and your index finger. As we saw in Chapter 3, people exposed to lots of androgens (the precursor to testosterone) during their fetal growth tend to have ring fingers longer than their index fingers; those of you exposed as a fetus to more estrogens have relatively longer index fingers. You may think you will be able to simply look at your hands and see at a glance which of your fingers is longer, but you'll want to measure with your small ruler—because the palm is round, the index finger can often look longer than it is. Be sure to measure from the bottom of the crease between each finger and the middle finger—don't measure the thumb side of your index finger or you may start your measurement too far down.

### How to do it

Ideally, your ruler has a metric scale and you can easily see the differences in length, but even if you're looking at a scale in inches and have to use fractions, you should be able to get a good enough measurement for our purposes. Anything within 2–3 millimeters is good enough.

The following illustration should make the process even clearer. Remember that D2 stands for second digit, or index finger, while D4 stands for fourth digit, or ring finger.

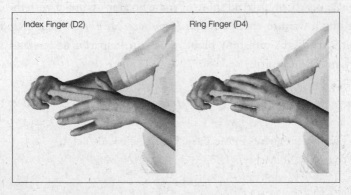

Index Finger (D2)          Ring Finger (D4)

What to do if the index and ring finger lengths are equal? No problem: "Ties" (equal measurements) always go to the index finger (D2), so if an index finger and ring finger are the same length, record the index finger as being the longer.

#### Record Your Index-Finger and Ring-Finger Measurements

| Right Hand: | D2 (Index Finger):<br>☐ Longer | D4 (Ring Finger):<br>☐ Longer |
|---|---|---|
| Left Hand: | D2 (Index Finger):<br>☐ Longer | D4 (Ring Finger):<br>☐ Longer |

If you've decided to do the Basic GenoType Calculator, that's it! Now move on to the next chapter and begin learning the techniques you will need to strength-test your GenoType.

# The Intermediate GenoType Calculator

The Intermediate GenoType Test uses the so-called classic genes—the single-gene measurements that tell us so much about our genetic heritage. Those of you who have read my Blood Type books are already familiar with how much information is revealed simply by knowing whether you're Type A, B, AB, or O. (And those of you who aren't familiar with the concept might want to check out *Eat Right for Your Type*, which includes a wealth of insight into how blood type can help you understand your body, mind, and spirit.)

*To perform the tests necessary for the Intermediate GenoType Calculator, you will need:*
- The measurements in the Basic GenoType Calculator.
- Your ABO blood type.

## Testing for ABO Blood Type

You may already know your ABO blood type. If not, you can:

- Call your doctor's office.
- Test yourself (see Resources to find out where to buy a kit).
- Donate blood—they'll test you at the same time and tell you the results.

Since some of the GenoTypes are linked to certain blood types, knowing your blood type is crucial to achieving the next level of accuracy. (The exception is, of course, our idiosyncratic GenoType 4 Explorers, the Swiss Army knife GenoType who contains a little of everything.)

### Record Your ABO Blood Type

| Your ABO Type (A,B,AB, O): |
| --- |

If you've decided to do the Intermediate GenoType Calculator, that's it! Now move on to the next chapter and begin learning the techniques you will need to strength-test your GenoType.

## The Advanced GenoType Calculator

If you are planning to use the Advanced GenoType Calculator, you've got all the data that you need to make the most sophisticated determination of your GenoType. In addition to measurements and blood types, you will now include your secretor status in the calculations. This gene has important effects on immune reactivity and metabolic thriftiness and can help distinguish subtle differences between the GenoTypes.

*To perform the tests necessary for the Advanced GenoType Calculator, you will need:*

- The measurements in the Basic GenoType Calculator.
- Your ABO and Rh blood types.
- Your secretor status (whether you are a "secretor" or a "non-secretor").

## Testing for Rh Blood Type

The Rh blood test is included in standard ABO blood typing panels and home test kits. If you already have done your blood type test, you'll have this information. Record it below. If not, get yourself tested for your blood type (see page 63) and record your results.

### Record Your Rh Blood Type

Your Rh Type (–, +):

## Testing for Secretor Status

Secretor status is closely linked to your ABO blood type, since the secretor gene controls whether or not you "secrete" your ABO blood type antigen in your body secretors. Whereas all of us could be blood-typed from a slide of blood, about 85 percent of us secrete our blood type in a free form that is pumped out into our body secretions, sweat, semen, mucus, saliva, and so on. These people are called *secretors,* and if we wanted to, we could probably blood-type them from a saliva or semen sample as well. About 15 percent of the world is missing the secretor allele, and not surprisingly, they are called *non-secretors.*

Secretor status has important effects on our metabolism and immune function. Non-secretors, for example, have lower levels of fat-busting enzymes and a tendency for inflammatory problems such as arthritis.

If you already know your secretor status, record it below. If you would like to use the Advanced GenoType Calculator and need to determine your secretor status, please refer to the Resources section in the back of the book. The test, which requires a saliva sample, is easy to do, and you'll get your results in about three weeks.

### Record Your Secretor Status

| ☐ Secretor | ☐ Non-secretor |
|---|---|

That's it! Now move on to the next chapter and begin learning the techniques you will need to strength-test your GenoType.

# CHAPTER FIVE

# Strength-Testing Your GenoType

**E**ach of the tests and measurements described in Chapter 4 will help us to distinguish one GenoType from another. The measurements and tests in this chapter are different. They're designed to help us understand how much you represent that GenoType as part of its total picture—how strong that GenoType resides in you. In science, this is called "goodness of fit."

Because of this, everyone who is calculating their GenoType does these tests. Regardless of whether you are doing the simple Basic Geno-Type Calculator or even the Advanced GenoType Calculator, you'll need these findings to see just how well you fit your GenoType description.

Everything you need to know about how to complete this chapter is right here. However, I should warn you that some of these questions may have you shaking your head, wondering how such odd bits of information could ever reveal anything useful. You can find a full explanation of the science behind every question in Chapter 3, and an overview of why this science matters in Chapters 1 and 2.

If you don't like the idea of collecting these items now, you can also breeze through this chapter and answer the easy questions first, then take the remaining measurements one by one. Do whatever seems easiest. And if there's any item you simply can't answer, just skip the question. There's no penalty for not knowing. The test simply becomes more accurate as you answer more questions.

So to recapitulate, here's what you've done so far, what you will do in this chapter, and what you will be doing in the next chapter:

- You've decided which calculator (Basic, Intermediate, or Advanced) works for you.
- You've done the measurements and tests required for that calculator.
- In this chapter, you'll take a few more measurements that you need and record and analyze your fingerprints.
- In the next chapter, you'll use the calculator you've chosen to determine your GenoType and use the information you've collected in this chapter to see how closely you reflect the more unique characteristics of your GenoType.
- Then you'll move on to the next part of the book and read the profile of your GenoType.
- Finally, you'll begin putting your GenoType information to work by using the dietary, exercise, and supplement information for your GenoType.

## What You'll Need to Strength-Test for Your GenoType

- Thirty minutes, with a friend present for most of that, to help you measure or act as an impartial observer.
- A soft tape measure, the kind you can use to measure your waist and hips.
- A protractor, masking tape, and crayon or water-based marker (you may be able to do without these).
- An ink pad, and some "bright white" laser-printer paper for taking fingerprints (and some alcohol and a rag for cleaning your hands afterward).
  - OR a fingerprint kit (see Resources for where you can buy one).
  - OR some time to go down to the local police station and for a small fee get fingerprinted.

- PROP taster strips (see Resources).
- Answers to some questions about you, your parents, and your grand-parents.

## Your Personal and Family History

Take a moment to answer the following questions. Every question has a "yes" or "no" answer. If you don't know the answer, just skip the question.

**Question #1: Are you sensitive to caffeine? Would a cup of coffee in the evening keep you wide awake at night or give you heart palpitations?**

My guess is that most people reading this question will have a strong gut reaction, either "Oh, my goodness, *yes!*" or "Well, coffee makes me a lit-tle jumpy, but it's not that big a deal." If you're really not sure, have a strong cup of coffee or perhaps some caffeinated green or black tea about two hours before your normal bedtime. (If the very idea makes you cringe, stop right there—you're sensitive.) I advise you not to try this the night before a workday.

**Question #2: Among you, your parents, grandparents, and siblings, have there been two or more instances of clinical depression or cognitive dysfunc-tion, such as Alzheimer's disease?**

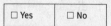

If you can count two or more instances, put "yes." It doesn't matter whether there are two, three, or more—two is the cutoff number, so you don't need a complete family history. And if you aren't sure, just skip the question. I want to caution you, though, that "depression" doesn't mean moodiness or sadness, but people who are (or should have been) under medication for depression, bipolar disorder, or obsessive-compulsive dis-order (OCD)—people whose functioning was seriously impaired by their condition, at least some of the time.

**Question #3: Among you, your parents, grandparents, and siblings, have there been two or more instances of heart disease, stroke, or diabetes?**

☐ Yes          ☐ No

**Question #4: Among you, your parents, grandparents, and siblings, have there been two or more instances of cancer?**

☐ Yes          ☐ No

Again, if you don't know the answers, just move on. But these are questions whose answers you really should know quite apart from this test, so please try to find out. It's definitely information that your physician should have!

**Question #5: Among you, your parents, grandparents, and siblings, have there been two or more instances of autoimmune disease?**

☐ Yes          ☐ No

Some common autoimmune diseases are rheumatoid arthritis, scleroderma, lupus, nephritis, colitis, and Crohn's disease. Allergies are also a form of autoimmune disease, so check the box if you or a close family member has severe reactions to pollen, mold, or environmental triggers. Don't include food allergies, which is a kind of catchall phrase that often includes "food sensitivities," an entirely different issue. And, when you're considering arthritis, make sure that the condition you're considering is rheumatoid arthritis, an inflammatory systemwide condition, and not osteoarthritis, which affects only the joints.

OK, the interrogation is over. We don't need every single bit of health information on you, just the answers to some key questions that will let us distinguish between aspects of the different GenoTypes. To find out what your fingerprints can tell you, read on.

## Fingerprints: A Prenatal Road Map

As we saw in Chapter 3, your fingerprints contain all sorts of clues about your prenatal and epigenetic heritage. Like the rings around a tree trunk, they are a sensitive record of your embryonic life and they can often indicate disorders you may need to watch out for—and for that reason, they're one of the best ways of figuring out your GenoType.

When it comes to getting your fingerprint data, you've got lots of choices. If you want to go down to your local police station, they'll fingerprint you for a small fee and let you take the prints home. This is quite common, since many jobs that require a financial bond and many license applications require that you submit a copy of your fingerprints. If you're the one who initiates the fingerprinting, they won't keep your prints on file—you'll be given all the results to take with you.

You can also send away for a GenoType testing kit, which includes a fingerprint kit. (See Resources for more information.)

Or you can simply take your own fingerprints, which is actually quite easy.

### How to do it

- Set out an inkpad and "bright white" laser-printer paper or some other paper with a smooth finish. Don't use cheap photocopy paper or high-quality bond; both finishes will be too rough or textured. The paper has to look polished and smooth.
- Wash your hands and dry them thoroughly. If you're concerned about perspiration, rub a bit of alcohol on your fingertips and then let them dry.
- Bring the inkpad and your paper right to the edge of the table. You want them as close to you as possible. The edge of the paper, in particular, should line right up with the edge of the table.
- Hold the inkpad steady with the hand you're not fingerprinting. Roll your finger lightly in the ink, from the inside (the side that's closest to your body) to the outside. (Do the opposite with your thumb—roll it from the outside in.) Don't press too hard. Be sure, though, to get ink down below the first crease. You're not interested in your fingertips, but rather the lower part of your finger, right down to the crease.

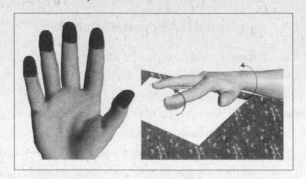

*Good technique leads to great fingerprints!*

- Press your inky finger lightly onto the paper. Again, roll it lightly, from inside to outside (except for your thumb, which you should roll from outside to inside). And again, don't press too hard. Let the ink do the work.
- If you're not used to fingerprinting yourself, you may need a few tries to get a good print, so be patient. If you have low fingerprint ridges (I'll explain in a minute), getting a good print becomes somewhat harder. Make sure you've gotten enough ink on your finger, but not too much. Not enough means your prints are too pale to be read; too much means they're all dark and blurry.
- Label your prints as soon as you take them, especially if you need more than one try to get them right. They'll all look alike after a while, so labeling is crucial. The standard system is to number the fingers from 1 to 5, with 1 being the thumb. Make sure you note which are from your left hand and which from your right.

## Interpreting your results

At first glance, your fingerprint patterns may seem awfully random. What you'll need to do, therefore, is to look for what's known as a ***pattern area***—the area within which you can actually see the pattern. Pattern areas are surrounded by ***type lines,*** ridges that basically outline the pattern.

Two landmarks can help you locate patterns as well. The ***triradius,*** or the ***delta,*** is a little triangle that's kind of an island in the midst of two flowing lines. The ***core*** is the center of the fingerprint pattern.

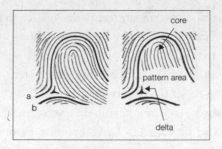

Once you've identified the pattern area, you'll be looking for three basic patterns: *arches, loops,* and *whorls.* You probably won't have all three patterns, but you'll almost certainly have at least one. You're going to want to count how many of each you have, and to notice some other variations as well, so you'll need to be able to identify each type.

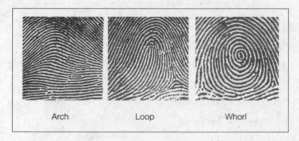

Arch        Loop        Whorl

- *Arches* are the simplest pattern—a succession of parallel ridges that look like a little hill. Sometimes the arches rise gently from either side; sometimes they form a peak in the center. Either way, for our purposes, that pattern counts as an arch.
- *Loops* are the most common pattern. They look like little lassos. The most common kind open toward your pinkie finger and are called *ulnar loop*s (the *ulna* is the bone on the outside of your arm). When they open toward your thumb, they're called *radial loops* (the radius is the bone on the inside of your arm). Radial loops are somewhat rare, but you may have such a loop on one or both index fingers. The type of loop doesn't matter for our purposes, only whether it is a loop or not.
- *Whorls* are series of concentric rings. They might be spiral, oval, circular, or anything else that looks round.

The best way to detect fingerprint patterns is to step back a bit and look at the whole pattern area; try not to get too hung up on the details. I sometimes say that it's like looking at modern art—or like those optical illusions where an unexpected figure emerges out of other images. Don't look too hard—let the pattern come to you.

You might discover a fingerprint that you can't categorize easily because it seems to be a blend of two or more patterns. In technical terms, that's called a *composite*. If you find one, count the deltas. One delta and you can call it a loop. Two deltas and you can call it a whorl. No deltas and you can call it an arch. If you absolutely can't categorize it, just let it go. Do the best you can with the data you have left.

## Record Your Fingerprint Pattern Results

|  | Thumb | Index | Middle | Ring | Pinkie |
|---|---|---|---|---|---|
| Right Hand | ☐ Loop<br>☐ Whorl<br>☐ Arch | ☐ Loop<br>☐ Whorl<br>☐ Arch | ☐ Loop<br>☐ Whorl<br>☐ Arch | ☐ Loop<br>☐ Whorl<br>☐ Arch | ☐ Loop<br>☐ Whorl<br>☐ Arch |
| Left Hand | ☐ Loop<br>☐ Whorl<br>☐ Arch | ☐ Loop<br>☐ Whorl<br>☐ Arch | ☐ Loop<br>☐ Whorl<br>☐ Arch | ☐ Loop<br>☐ Whorl<br>☐ Arch | ☐ Loop<br>☐ Whorl<br>☐ Arch |

## White Lines

Your fingerprint patterns are pretty much fixed by the time you're born (which is why they're so useful to law enforcement). What does change, though, is the height of your fingerprint *ridges*, the texture of your fingers that give you fingerprints in the first place. Ridge height is dynamic, and it's often a very strong indicator of what's going on in your digestive system. Low ridges often indicate disruption in the lining of your intestinal tract or some other type of digestive problem. They could also indicate intolerance to gluten (found in wheat) or sensitivities to lectins (found in grains), as well as suggesting celiac disease (related to gluten intolerance) and "leaky gut," in which bacteria that belong in your stomach migrate into your intestinal tract as well. By contrast, good ridge height usually indicates a healthy digestive tract.

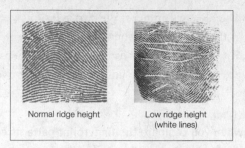

Normal ridge height       Low ridge height
                          (white lines)

If your fingerprint ridges are worn, you're likely to see a pattern of white lines among your fingerprints—secondary creases on your fingers that become visible when your ridges are low.

Research dating back to the early 1970s shows a correlation between the appearance of white lines and the incidence of celiac disease. Typically, the number of white lines increases with age as gut integrity continues to deteriorate. In many cases, these white lines begin to vanish with the maintenance of a gluten-free diet. Some researchers even believe that white lines are a useful indicator of a person's response to diet therapy, although complete improvement of the fingerprints might take as long as two years.

If you've noticed lots of white lines in your fingerprints, you might want to talk with a naturopathic physician to learn more about your digestive health. And you will certainly want to follow the carbohydrate prescriptions in your GenoType Diet, which can make a world of difference in correcting digestive problems, restoring gut integrity, and rebalancing stomach and intestinal bacteria.

**Question #6: Do you have white lines throughout your fingerprint patterns?**

☐ Yes          ☐ No

## Symmetries

As we saw in Chapter 3, symmetry usually indicates a stable prenatal environment, while asymmetry suggests that the fetus was subject to stress. Generally, the more asymmetrical you are, the more stress you experienced in the womb. That's because the left and right sides of your body develop separately. In a stable environment, the two sides follow the same logic and develop in the same way. When stress interferes with the fetus's development, it tends to do so unevenly—and asymmetry is the result. That's why we're interested in the answers to the first two questions in this section:

**Question #7: When you compare the fingerprint patterns of your left and right hands to each other, at least four out of five fingers DO match.**

**Question #8: When you compare the fingerprint patterns of your left and right hands to each other, at least three out of five fingers DON'T match.**

Left-handedness and ambidexterity tend to correlate to hormonal shifts during pregnancies, which accounts for Question #9:

**Question #9: Are you are left-handed or ambidextrous?**

| ☐ Yes | ☐ No |
|-------|------|

As we saw in Chapter 3, none of these patterns *causes* anything to happen in your body. They're simply *records* of events that have taken place—events that might have profound effects upon your health and vitality.

## Biometrics: Measuring Your Body's Shape and Size

Biometrics is literally the measure of living things. In this case, we're measuring your ***morphology,*** a word that means the study of shapes. The way your body is shaped reveals quite a bit about your hormones and metabolism, and gives us a great deal of insight into how your prenatal experience has shaped your responses to the world. As always, you can find the scientific explanations for these measurements in Chapter 3, so let's get right down to the actual GenoType Test.

## Leg Opening

**Question #10: With your ankles together, is there a SMALL opening between the upper legs at the area of the knees, or do the knees touch?**

**Question #11: With your ankles together, is there a LARGE opening between the upper legs at the area of the knees?**

The answer to these questions couldn't be simpler to find out. Just look at your bare legs in a full-length mirror and notice the space that's created between your thighs when your feet are placed together, touching gently at the ankles. You can use the drawing below to help you decide whether the opening between your thighs is small or large—and you also might want to get a friend's opinion.

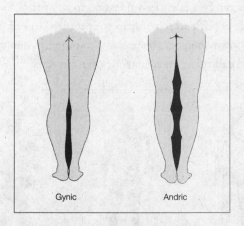

*Upper-leg space: (left) narrow opening; (right) wide opening*

## Tendons and Sinews

**Question#12: When you relax your arm and look at the skin around your wrists, can you see the outline of your tendons?**

| ☐ Yes | ☐ No |
|---|---|

The answer is a keynote for Teacher GenoTypes, who tend to have sinewy bodies with visible tendons. Not all of them do, however, and if you're overweight, you may not be able to view your tendons in any case. If you're not sure about this issue, just leave it blank.

# Body Type

As we saw in Chapter 3, your body type—also known as your *somatotype*—can tell us quite a bit about your metabolism and your GenoType. For example, the rounded body type is also a keynote for the Gatherer GenoType, whereas Hunter GenoTypes tend toward a more lanky body type.

Again, this can be a fairly easy question to answer, although I'm going to suggest you get a friend to confirm your results. In our "thin is in" culture, lots of women consider themselves "rounded" when virtually everyone else would call them "muscular" or even "lanky."

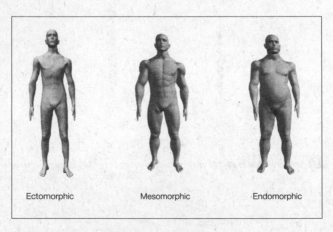

Ectomorphic        Mesomorphic        Endomorphic

*The three basic body types*

Take a look at these body types and decide which one best describes you. Then have your friend do the same. If you both agree, you're done. If you are still unclear about which body type best describes you, try this simple test:

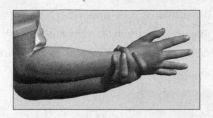

*The wrist circumference test.*

Circle your wrist with the thumb and middle finger of the other hand (see photo). Which of the following did you notice?

- Your middle finger and thumb did not touch. You are large-framed and most likely an endomorph.
- Your middle finger and thumb just touch. You are almost certainly a mesomorph.
- Your middle finger and thumb overlap. You are small-framed and most likely an ectomorph.

*Recording Your Results*
- ☐ You have a more ROUNDED body type and are most likely large-framed (endomorph).
- ☐ You have a more MUSCULAR body type (mesomorph).
- ☐ You have a more LANKY body type and are probably small-framed (ectomorph).
- ☐ You have a ROUNDED, MUSCULAR body type (meso-endomorph).
- ☐ You have a more MUSCULAR, LANKY body type (meso-ectomorph).

## Teeth Patterns

As we saw in Chapter 3, tooth shoveling—a scooped-out shape in the back of your two front teeth—may indicate ancestors who were used to eating meat. Accordingly, it's a keynote for GenoType 1 Hunters and a lesser indicator for GenoType 2 Gatherers and GenoType 4 Explorers— who also do better on a carnivore-type diet.

If shoveling suggests meat-eating, an extra cusp—known as Carabelli's cusp—on the inside of your upper front molar suggests an agrarian diet that requires the grinding of grains and vegetables.

Take a look at the following drawing to get an idea of the areas you will need to be looking at.

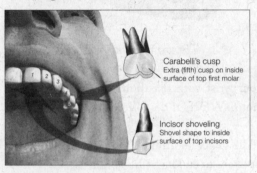

*Tooth shoveling and Carabelli's cusp*

Of course, if you've had dental work done on any of the teeth we're interested in, you won't be able to answer these questions. You may want to enlist a friend to have a look around your mouth with a pocket flashlight. Don't worry—there are plenty of other questions!

**Question #13: Do you have SPADE-LIKE front incisor teeth (shoveling)?**

☐ Yes      ☐ No

**Question #14: Do you have the EXTRA CUSP on the inside of your front molar?**

☐ Yes      ☐ No

## Jaw Shape

Several GenoTypes are distinguished by their unique jaw shapes, which you will probably be able to determine simply by looking in the mirror, using the following illustration to guide you:

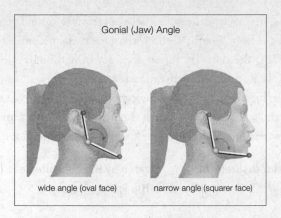

*(Left) Wide gonial angle. (Right) Narrow gonial angle.*

Two examples of different jaw angles are shown in the above graphic. On the left is a wide jaw (angle greater than 125 degrees) and on the right is a narrow jaw (angle less than or equal to 125 degrees).

If you are unsure, you can find or take a photo of yourself in profile. Mark dots on the three landmarks, draw lines connecting them, and measure the angle with a protractor. If the angle you measure is 125 degrees or less, you have a square jaw (a narrow gonial angle). If your jaw angle is greater than 125 degrees, you have an almond-shaped jaw (a wide gonial angle). Now you are ready to record your results and move on to the next measurement.

*Recording Your Results*
☐   You have a more SQUARE jaw (narrow gonial angle).
☐   You have a more ALMOND-shaped jaw (wide gonial).

## Waist-to-Hip Ratio

As we saw in Chapter 3, comparing your waist to your hip ratio reveals quite a bit about your health and your metabolism.

### How to do it
By yourself or with the help of a friend, wrap a soft tape measure around your waist at its narrowest point, just above the belly button.

Make sure your tape measure is level all the way around your body and parallel to the floor. Gently tighten the tape measure until it's snug but not depressing the skin. Try for an accuracy of within 10 millimeters, or ⅛ of an inch. Record your waist measurement.

For your hips, repeat the procedure, choosing the widest part of your hipbones. Record your hip measurement. Then divide your waste measurement by your hip measurement. (Now you know why you needed the calculator!)

**Waist Measurement: _____ divided by Hip Measurement: _____**
**= Waist-to-Hip Ratio _____**

Now let's classify it:

*Men, under 50 years of age:*
- High (your waist-to-hip ratio is over .96)
- Average (your waist-to-hip ratio is .96–.91)
- Ideal (your waist-to-hip ratio is under .91)

*Women, under 50 years of age:*
- High (your waist-to-hip ratio is over .79)
- Average (your waist-to-hip ratio is .79–.71)
- Ideal (your waist-to-hip ratio is under .71)

*Men, ages 50 and older:*
- High (your waist-to-hip ratio is over .99)
- Average (your waist-to-hip ratio is .99–.93)
- Ideal (your waist-to-hip ratio is under .93)

*Women, ages 50 and older:*
- High (your waist-to-hip ratio is over .84)
- Average (your waist-to-hip ratio is .85–.75)
- Ideal (your waist-to-hip ratio is under .75)

*Recording Your Results*
We are interested only in whether your waist-to-hip ratio is exceptionally high or ideal, so if your results are "average," go ahead and move on to the final measurement. Otherwise record your results on the next page.

☐  You have a HIGH waist-to-hip ratio.

☐  You have an IDEAL waist-to-hip ratio.

## Head Shape

You're almost there! Of all things, your final set of questions in the strength-testing section concerns your head shape.

As I said in Chapter 3, head shape seems to be one of the more recent changes we humans have undergone. Up to the Middle Ages, our heads seem to have been gradually widening to more of a "blockhead" shape. From the Middle Ages onward, our average head shape appears to have been elongating and narrowing, in combination with the gradual increase in our height. Because these changes probably reflect differences in maternal nutrition, they correlate to certain GenoTypes, especially the elongated head, which is a keynote to Warriors. The squarish head is also somewhat of a characteristic of Explorer and Nomad.

There are three basic head shapes. Use the following illustration to help you determine your head shape:

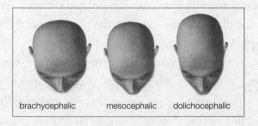

*The three basic head shapes*

Most people can get a good idea of their head shape by having a friend view the top of their head while seated. If you have long or wavy hair, you may want to wet it beforehand so that it lies flat. If you can't tell by sight alone and want to measure, get out that soft tape measure again and enlist the help of a friend. You'll be measuring what anthropologists call the "cephalic index."

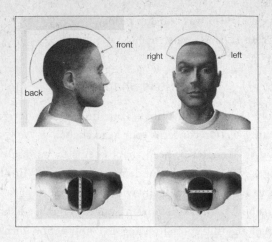

*(Top) Measuring for the cephalic index*
*(Bottom left) Making the back-to-front measurement*
*(Bottom right) Making the side-to-side measurement*

### How to do it

- Referring to the diagram above, measure the length of the head from the highest point at the midline of your head—the crown of your skull (**front** in the illustration)—down to the bony bump at the base of your skull (**back** in the illustration). Record the length.

- Measure the width of your head by running a tape measure over the top of your skull at its widest point. (See the illustration for guidance—you're looking for points labeled **left** and **right** on both sides of your skull.) Record the width.

- Divide your head width by the head length and multiply by 100. That's your cephalic index:

**Head Width: _____ divided by Head Length: _____ multiplied by 100**
**= Cephalic Index: _____.**

*And here's what your cephalic index tells you:*
- If your cephalic index is less than 75, you've got a long, narrow skull (as seen from above). You're *dolichocephalic,* which is Greek for "long-headed."

- If your cephalic index is more than 80, you've got a short, broad head (as seen from above), which makes you *brachycephalic*, or "short-headed."
- In the middle lie those whose cephalic index falls between 75 and 80. If you're in this group, you're *mesocephalic*, or "medium-headed."

*Recording Your Results*
We are interested only in whether you are long-headed or short-headed, so if your results are "normal" (i.e., "medium-headed"), you're done!
   Otherwise, record your results below:
☐ You have an ELONGATED head shape (dolichocephalic).
☐ You have a BROAD, SHORT head shape (brachycephalic).

## Testing for PROP Taster Status

As we saw in Chapter 3, PROP tasters have a specific gene that enables them to taste a chemical known as propylthiouracil (PROP), similar to many of the compounds found in cruciferous vegetables such as broccoli, cabbage, and cauliflowers. The ability or inability to taste PROP can tell us a lot about the body's metabolism and can help determine many of the finer distinctions among the GenoTypes.

### How to do it

If you've gotten a GenoType Test kit (see Resources), you've already gotten a PROP test strip and instructions on how to use it. You can also buy a PROP test strip at educational-supply outlets that sell supplies for high school biology classes. Make sure to get a control strip, too. You need the control strip to make sure you really can taste the PROP and are not just imagining it. Mix up the strips, or have a friend try to fool you by having you taste each strip with your eyes closed.

*Interpreting your results*
- The control strip should taste "blank."
- If the PROP strip also tastes "blank," you are a "non-taster."
- If the PROP strip tastes bitter, you are a "taster."

- If the PROP strip strongly repels you (instead of being simply a bad taste), you're a "super-taster."

*Record Your PROP Taster Status*
☐ Taster
☐ Non-taster
☐ Super-taster

## Moving On

And now you're done! Soon you will be able to access a whole new world of health, vitality, and optimal weight. And you may understand your physical, mental, and emotional self in a whole new way.

Take this opportunity to look back over this chapter and the previous one to make sure you've answered all the questions you can, checking with friends or loved ones about any question that needs a fresh, objective eye. Make sure, for example, that the body type you've identified as your own is one that your friends would agree with. Correct anything that doesn't seem right to you. Then pick up your calculator and prepare to add it all up. Take one last look at your jaw angle or your head shape if you still have any doubt about how to describe them.

Then turn the page to find out how to score the results to your Geno-Type Test.

# CHAPTER SIX

# Ready, Set, Go!
# Calculating Your GenoType

Y ou know the science. You've gotten answers to the Five Basic Questions about your personal and family history. You've done your measurements and taken the tests. Now it's time to get out the GenoType Calculator and discover which one of our six GenoTypes you are.

OK, let's do it!

## The Basic GenoType Calculator

*What You Need*

- Leg- and torso-length measurements, as outlined in Chapter 4.
- Index-finger and ring-figure measurements for both hands, as outlined in Chapter 4.
- Any measurements, fingerprints, or observations you will use to strength-test your GenoType, as outlined in Chapter 5.

Running the Basic GenoType Calculator is very simple. Just do the simple measurements outlined in Chapter 4, go to the appropriate male/female lookup table, and find the four GenoTypes the calculator is telling you to "strength-test" against one another. (Remember, a "tie"—equal measurement—goes to the torso, lower legs, and index fingers.) Then move on to the section called "Strength-Testing Your GenoType" at the end of this chapter. There you'll take the information that you accumulated in Chapter 5 and use it to test the four GenoTypes against one another. The GenoType with the highest score is your GenoType.

*Basic GenoType Calculator: Table 1*

## Your TORSO is LONGER than or equal in length to your LEGS

| And . . . | And . . . | Strength-Test These GenoTypes |
|---|---|---|
| Your upper leg is longer than your lower leg | Your index fingers are longer than your ring fingers on both hands | GT2 GATHERER<br>GT3 TEACHER<br>GT4 EXPLORER<br>GT6 NOMAD |
| | Your ring fingers are longer than your index fingers on both hands | GT1 HUNTER<br>GT3 TEACHER<br>GT4 EXPLORER<br>GT6 NOMAD |
| | Your index finger is longer on one hand and your ring finger is longer on the other | GT2 GATHERER<br>GT3 TEACHER<br>GT4 EXPLORER<br>GT5 WARRIOR |
| Your lower leg is longer than or equal to your upper leg | Your index fingers are longer than your ring fingers on both hands | GT2 GATHERER<br>GT4 EXPLORER<br>GT5 WARRIOR<br>GT6 NOMAD |
| | Your ring fingers are longer than your index fingers on both hands | GT1 HUNTER<br>GT3 TEACHER<br>GT4 EXPLORER<br>GT6 NOMAD |
| | Your index finger is longer on one hand and your ring finger is longer on the other | GT2 GATHERER<br>GT3 TEACHER<br>GT4 EXPLORER<br>GT6 NOMAD |

*Basic GenoType Calculator: Table 2*

## Your LEGS are LONGER in length than your TORSO

| And ... | And ... | Strength-Test These GenoTypes |
|---|---|---|
| Your upper leg is longer than your lower leg | Your index fingers are longer than your ring fingers on both hands | GT2 GATHERER<br>GT4 EXPLORER<br>GT5 WARRIOR<br>GT6 NOMAD |
| | Your ring fingers are longer than your index fingers on both hands | GT1 HUNTER<br>GT4 EXPLORER<br>GT5 WARRIOR<br>GT6 NOMAD |
| | Your index finger is longer on one hand and your ring finger is longer on the other | GT1 HUNTER<br>GT2 GATHERER<br>GT3 TEACHER<br>GT4 EXPLORER |
| Your lower leg is longer than or equal to your upper leg | Your index fingers are longer than your ring fingers on both hands | GT2 GATHERER<br>GT4 EXPLORER<br>GT5 WARRIOR<br>GT6 NOMAD |
| | Your ring fingers are longer than your index fingers on both hands | GT1 HUNTER<br>GT3 TEACHER<br>GT5 WARRIOR<br>GT6 NOMAD |
| | Your index finger is longer on one hand and your ring finger is longer on the other | GT1 HUNTER<br>GT3 TEACHER<br>GT5 WARRIOR<br>GT6 NOMAD |

I suppose it can't get any simpler than this!

The Basic GenoType Calculator is just that: basic. It will get you roughly into the right ballpark and that is about it. I think of the Basic GenoType Calculator as much like a "genetic finger painting." Have you ever looked at a child's finger painting? Although most are simplistic, more often than not you're able to pinpoint the subject of the painting— a house, or a car, or a smiling person—and you might be surprised by the considerable nuance that children can often put into their drawings. However, since the Basic GenoType Calculator uses only two measurements, it is certainly possible that your GenoType will change as you move into the more detailed calculators.

Simple as it is, the Basic GenoType Calculator is quite powerful and its results are heads above any of the "one-size-fits-all" diets out there. We should not underestimate the power of these little tables. They do a pretty good job of determining some basic epigenetic influences, and factor in some important details about your levels of symmetry, the amount of sex hormones that you encountered while in utero, and the levels of growth factors that you experienced as a young child.

The idea behind the Basic GenoType Calculator was to simply get you "up and moving" with the GenoType Diet. In time you'll want to add more data about yourself, such as your blood type, and you can move up to the Intermediate GenoType Calculator.

## The Intermediate GenoType Calculator

*What You Need*
- Leg- and torso-length measurements, as outlined in Chapter 4.
- Index-finger and ring-finger measurements for both hands, as outlined in Chapter 4.
- ABO blood type, as outlined in Chapter 4.
- Any measurements, fingerprints, or observations you will use to strength-test your GenoType, as outlined in Chapter 5.

The Intermediate GenoType Calculator integrates one of your genetic markers—your ABO blood type.

If you don't know your ABO blood type, a simple and inexpensive home test can be ordered from sources listed in Resources. Just like the Basic GenoType Calculator, you will perform the two basic measurements, and again, there are two lookup tables, one for the folks with torsos longer than their legs, and the other for those with legs longer than their torsos.

If we were to stick with our painting analogy, the Intermediate Geno-Type Calculator moves the accuracy of your GenoType determination from the finger-painting level on up to the level of the Old Masters, since just like a Rembrandt oil painting, the blood-type genetics now allow for a more discreet and delicate shading. Many readers may already have this information from having read my other books on blood types and diet.

To use the Intermediate GenoType Calculator, choose the lookup table based on torso-to-leg length and work through the table from left to right. Unlike the Basic Calculator, which asks you to strength-test four possible GenoTypes, the Intermediate Calculator narrows the possible choices down to two candidates and in special circumstances will even specify a single, winning GenoType.

After you have done the Intermediate GenoType Calculator, you can move on to the section at the end of the chapter called "Strength-Testing Your GenoType." There you'll take the information that you accumulated in Chapter 5 and use it to test the two GenoTypes against each other. The GenoType with the highest score is your GenoType. If the calculator has specified a single GenoType, you're done.

| Intermediate GenoType Calculator: Table 1 | | | |
|---|---|---|---|
| **Your TORSO is LONGER than or equal in length to your LEGS** | | | |
| *And . . .* | *And . . .* | *And You Are Blood Type . . .* | *GenoTypes to Strength-Test* |
| Your upper leg is longer than your lower leg | Your index fingers are longer than your ring fingers on both hands | A | GT3 TEACHER GT4 EXPLORER |
| | | AB | GT4 EXPLORER GT6 NOMAD |
| | | B | GT6 NOMAD GT4 EXPLORER |
| | | O | GT2 GATHERER |
| | Your ring fingers are longer than your index fingers on both hands | A | GT3 TEACHER GT4 EXPLORER |
| | | AB | GT4 EXPLORER GT6 NOMAD |
| | | B | GT4 EXPLORER GT6 NOMAD |
| | | O | GT1 HUNTER GT4 EXPLORER |
| | Your index finger is longer on one hand and your ring finger is longer on the other | A | GT3 TEACHER |
| | | AB | GT4 EXPLORER GT5 WARRIOR |
| | | B | GT2 GATHERER GT4 EXPLORER |
| | | O | GT2 GATHERER GT4 EXPLORER |

*(continued on next page)*

*Intermediate GenoType Calculator: Table 1 (continued)*

## Your TORSO is LONGER than or equal in length to your LEGS

| And . . . | And . . . | And You Are Blood Type . . . | GenoTypes to Strength-Test |
|---|---|---|---|
| Your lower leg is longer than or equal to your upper leg | Your index fingers are longer than your ring fingers on both hands | A | GT4 EXPLORER GT5 WARRIOR |
| | | AB | GT5 WARRIOR GT6 NOMAD |
| | | B | GT2 GATHERER GT6 NOMAD |
| | | O | GT2 GATHERER GT4 EXPLORER |
| | Your ring fingers are longer on both hands | A | GT3 TEACHER GT4 EXPLORER |
| | | AB | GT3 TEACHER GT4 EXPLORER |
| | | B | GT6 NOMAD GT4 EXPLORER |
| | | O | GT1 HUNTER GT4 EXPLORER |
| | Your index finger is longer on one hand and your ring finger is longer on the other | A | GT3 TEACHER |
| | | AB | GT3 TEACHER |
| | | B | GT2 GATHERER GT6 NOMAD |
| | | O | GT2 GATHERER GT4 EXPLORER |

| Intermediate GenoType Calculator: Table 2 | | | |
|---|---|---|---|
| Your LEGS are LONGER in length than your TORSO | | | |
| And . . . | And . . . | And You Are Blood Type . . . | GenoTypes to Strength-Test |
| Your upper leg is longer than your lower leg | Your index fingers are longer than your ring fingers on both hands | A | GT4 EXPLORER GT5 WARRIOR |
| | | AB | GT5 WARRIOR GT6 NOMAD |
| | | B | GT2 GATHERER GT6 NOMAD |
| | | O | GT2 GATHERER |
| | Your ring fingers are longer than your index fingers on both hands | A | GT2 GATHERER GT5 WARRIOR |
| | | AB | GT4 EXPLORER GT5 WARRIOR |
| | | B | GT4 EXPLORER GT6 NOMAD |
| | | O | GT1 HUNTER GT4 EXPLORER |
| | Your index finger is longer on one hand and your ring finger is longer on the other | A | GT3 TEACHER |
| | | AB | GT3 TEACHER GT4 EXPLORER |
| | | B | GT2 GATHERER GT4 EXPLORER |
| | | O | GT1 HUNTER GT2 GATHERER |

(continued on next page)

*Intermediate GenoType Calculator: Table 2 (continued)*

## Your LEGS are LONGER in length than your TORSO

| And . . . | And . . . | And You Are Blood Type . . . | GenoTypes to Strength-Test |
|---|---|---|---|
| Your lower leg is longer than or equal to your upper leg | Your index fingers are longer than your ring fingers on both hands | A | GT5 WARRIOR |
| | | AB | GT5 WARRIOR GT6 NOMAD |
| | | B | GT2 GATHERER GT6 NOMAD |
| | | O | GT2 GATHERER GT4 EXPLORER |
| | Your ring fingers are longer than your index fingers on both hands | A | GT3 TEACHER GT5 WARRIOR |
| | | AB | GT5 WARRIOR GT6 NOMAD |
| | | B | GT6 NOMAD |
| | | O | GT1 HUNTER |
| | Your index finger is longer on one hand and your ring finger is longer on the other | A | GT3 TEACHER GT5 WARRIOR |
| | | AB | GT5 WARRIOR GT6 NOMAD |
| | | B | GT6 NOMAD |
| | | O | GT1 HUNTER |

Now, human nature being what it is, a few of you will no doubt want to compare the results of computing your GenoType with the Intermediate Calculator to the results from computing your GenoType with the Basic Calculator—especially when strength-testing the GenoTypes.

My advice is, don't do it! Here's why:

Let's say that you use the Basic Calculator and it tells you to strength-test four GenoTypes, and in your case the Gatherer GenoType strength-tests the highest. Then you order a home blood-typing test and find out that you are blood type A. Now, using the Intermediate Calculator, you discover that you are to instead strength-test only the Teacher and Explorer.

You may find yourself asking, "What gives? When I did the Basic Calculator, neither of these GenoTypes strength-tested as high as the Gatherer."

That indeed may be true, but you would be forgetting the effect of including that all-important piece of ABO blood-type data, which is the equivalent of adding perhaps 15 strength points to each of those two GenoTypes. By including ABO blood type, the Intermediate Calculator provides an additional dimension to its conclusions, a dimension that the Basic Calculator lacks.

## The Advanced GenoType Calculators

*What You Need*

- Leg- and torso-length measurements, as outlined in Chapter 4.
- Index-finger and ring-finger measurements for both hands, as outlined in Chapter 4.
- ABO blood type, as outlined in Chapter 4.
- Rhesus (Rh) blood type, as outlined in Chapter 4.
- Secretor status (optional), as outlined in Chapter 4.
- Any measurements, fingerprints, or observations you will use to strength-test your GenoType, as outlined in Chapter 5.

You've got all the data that you need to make the most sophisticated determination of your GenoType. In addition to measurements and ABO blood type, we will now include your Rh blood group and secretor status in the calculations. In some instances, your gender will also be factored into our calculations, so if gender is a distinguishing factor you will find two GenoTypes listed—one for women and the other for men.

To stick with our painting analogy, the Intermediate GenoType Calculators move the accuracy of your GenoType determination up from the finger-painting level, past the Old Masters level, to a point now that might be compared to a high-resolution satellite photograph.

As with the other calculators, the tests used for the Advanced Geno-Type Calculators are easy to do and quite inexpensive. You will need to know your ABO and Rh blood types and your secretor status to complete the Advanced GenoType Calculators. If you don't know your secretor status, make a note to get the submission test and run it on yourself. Both of these tests can be ordered from sources listed in Resources.

Because of the large number of data points you will be using to determine your GenoType, if the Advanced Calculator were one big table, it would be too large to print! So I've broken it down into four lookup tables. They are listed in the Appendix. Use the chart below to see which lookup table you should use to calculate your GenoType:

| Advanced GenoType Calculator Jump Table | | |
|---|---|---|
| If your . . . | And . . . | Use Look-up Table (Appendix, p. 289) |
| Torso is longer than or equal to your legs | Your upper leg is longer than your lower leg | #1 |
| Torso is longer than or equal to your legs | Your lower leg is longer than or equal to your upper leg | #2 |
| Legs are longer than your torso | Your upper leg is longer than your lower leg | #3 |
| Legs are longer than your torso | Your lower leg is longer than or equal to your upper leg | #4 |

The Advanced GenoType Calculator will identify a single GenoType, so there is no need to strength-test the GenoType. Just for fun, you can move on to the "Strength-Testing Your Genotype" section at the end of this chapter and see how well you fit your GenoType, or you can skip to the next chapter and meet your GenoType.

## Strength-Testing Your GenoType

How strong is your GenoType? Each of the six GenoTypes has its own unique characteristics, and how well you fit the "total picture" of your GenoType is a measurement of the strength of the effects that your GenoType exerts in your body.

Now it is time to use the information you collected in Chapter 5 to double-check your GenoType Calculator results. If you are using the Basic GenoType Calculator, you will be strength-testing the four suggested GenoTypes that the calculator recommended for you. The one with the highest score will be your GenoType.

If you are using the Intermediate GenoType Calculator, you will be strength-testing the two suggested GenoTypes that the calculator recommended for you. The one with the highest score will be your GenoType.

Strength-testing your GenoType could not be easier. Find the "Strength Meter" for the particular GenoType(s) you are interested in and check any of the boxes for statements that are true for you. Then add up the numbers to determine the strength points for that GenoType. Finally, check your results against the scorecard at the end of the chapter to see how closely you fit the description.

| The GT1 Hunter "Strength Meter" | |
|---|---|
| *Check any of the following statements that apply to you* | *Add these points to your total* |
| ☐ You have white lines on your fingerprints. | +5 |
| ☐ You are a PROP "super-taster." | +5 |
| ☐ Comparing left to right hands, four or more fingerprint patterns match. | +5 |
| ☐ You have spade-like front incisor teeth (shoveling). | +3 |
| ☐ You have a lanky body type (ectomorph). | +3 |
| ☐ There is a large opening between the upper legs. | +3 |
| ☐ You have a square-shaped jaw. | +3 |
| ☐ Between you, your parents, grandparents, and siblings, there have been two or more instances of autoimmune disease (lupus, rheumatoid arthritis, multiple sclerosis). | +3 |
| *Add your total* | |

| The GT2 Gatherer "Strength Meter" | |
| --- | --- |
| *Check any of the following statements that apply to you* | *Add these points to your total* |
| ☐ Comparing left to right hands, three or more fingerprint patterns don't match. | +5 |
| ☐ You are a PROP "non-taster." | +5 |
| ☐ Your skin looks "padded" even in areas where there is no fat tissue. | +5 |
| ☐ You have the extra molar cusp. | +3 |
| ☐ You have an almond-shaped jaw and face (wide gonial angle). | +3 |
| ☐ You have a rounded body type (endomorph) or a high waist-to-hip ratio. | +3 |
| ☐ You have a small opening between the upper legs or your knees touch. | +3 |
| ☐ Between you, your parents, grandparents, and siblings, there have been two or more instances of diabetes, stroke, or high blood pressure. | +3 |
| *Add your total* | |

| The GT3 Teacher "Strength Meter" | |
| --- | --- |
| *Check any of the following statements that apply to you* | *Add these points to your total* |
| ☐ You have five or more whorl-type fingerprint patterns. | +5 |
| ☐ You can see the sinews and tendons under the skin of your wrist. | +5 |
| ☐ You are a PROP "taster." | +5 |
| ☐ You have the extra molar cusp. | +3 |
| ☐ You have a square-shaped jaw and face (narrow gonial angle). | +3 |
| ☐ You have a lanky, muscular body type (meso-ectomorph) or an ideal waist-to-hip ratio. | +3 |
| ☐ There is a large opening between the upper legs. | +3 |
| ☐ Between you, your parents, grandparents, and siblings, there have been two or more instances of cancer. | +3 |
| *Add your total* | |

| The GT4 Explorer "Strength Meter" | |
|---|---|
| *Check any of the following statements that apply to you* | *Add these points to your total* |
| ☐  You are Rhesus-negative (Rh–). | +5 |
| ☐  You are a PROP "super-taster." | +5 |
| ☐  Are you sensitive to caffeine? Would a cup of coffee in the evening keep you awake at night? | +5 |
| ☐  You are left-handed or ambidextrous. | +3 |
| ☐  You have a square-shaped jaw and face (narrow gonial angle). | +3 |
| ☐  You have a muscular body type (mesomorph) or an ideal waist-to-hip ratio. | +3 |
| ☐  You have a broad, short head shape (brachycephalic). | +3 |
| ☐  Your index fingers have different fingerprint patterns. | +3 |
| *Add your total* | |

| The GT5 Warrior "Strength Meter" | |
|---|---|
| *Check any of the following statements that apply to you* | *Add these points to your total* |
| ☐ Your head shape is elongated (dolichocephalic). | +5 |
| ☐ You are a PROP "non-taster." | +5 |
| ☐ You have two or more arch-type fingerprint patterns. | +5 |
| ☐ You have a muscular to rounded body type (meso-endomorph) or a high waist-to-hip ratio. | +3 |
| ☐ You have an almond-shaped jaw and face (wide gonial angle). | +3 |
| ☐ You have the extra molar cusp. | +3 |
| ☐ Caffeinated drinks don't particularly bother you. | +3 |
| ☐ Between you, your parents, grandparents, and siblings, there have been two or more instances of diabetes, stroke, or heart disease | +3 |
| *Add your total* | |

| The GT6 Nomad "Strength Meter" | |
| --- | --- |
| *Check any of the following statements that apply to you* | *Add these points to your total* |
| ☐ You have white lines on your fingerprints. | +5 |
| ☐ You have eight or more loop-type fingerprint patterns. | +5 |
| ☐ You are a PROP "taster." | +5 |
| ☐ Comparing left to right hands, four or more fingerprint patterns match. | +3 |
| ☐ Your head shape is broad and short (brachycephalic). | +3 |
| ☐ You have spade-like front incisor teeth (shoveling). | +3 |
| ☐ You have an almond-shaped jaw and face (wide gonial angle). | +3 |
| ☐ Between you, your parents, grandparents, and siblings, there have been two or more instances of clinical depression or cognitive dysfunction such as Alzheimer's disease. | +3 |
| *Add your total* | |

*Interpreting the Results*

- If you scored above 20 points, you test VERY STRONG for your GenoType. Not only do you manifest the genetic and epigenetic characteristics of your GenoType, but you also show a great many of the expected physical manifestations of your GenoType.

- If you scored anywhere from 11 to 20 points, you test STRONG for your GenoType. You manifest the genetic and epigenetic characteristics of your GenoType and show many of the expected physical manifestations of your GenoType.

- If you scored anywhere from 5 to 10 points, you test POSITIVE for your GenoType. You manifest the genetic and epigenetic characteristics of your GenoType and show many of the expected physical manifestations of your GenoType.

- If you scored under 5 points, don't worry—you demonstrate the major epigenetic indicators for your GenoType, but some of the "hallmark" characteristics of your GenoType were not easy to determine. However, the epigenetic, life-altering actions of the GenoType Diet will still work as well as ever. Remember, Chewbacca doesn't look very much like Little John! If you have done the Basic GenoType Calculator, you may want to go ahead and determine your ABO blood type, and move to the Intermediate GenoType Calculator.

## The Whole Is Greater Than the Sum of Its Parts

There are no combination types: You belong to one GenoType and one GenoType only. The GenoTypes are a complete set of solutions; out of the six possibilities, only one will work the best for you.

Epigenetics has been likened to placing a marble at the top of a mountain and letting go. On its way downhill, the marble will make a series of choices as it enters valleys, pastures, and furrows. However, once it enters each valley or furrow there is no going back. The marble may notice other marbles rolling down the mountain and entering other valleys, and maybe even whistle and wave to them. Yet however close the two marbles are to each other physically, they lie in different valleys and thus will have different ending locations, different outcomes.

Remember our chat about archetypes? If Little John and Chewbacca were sitting next to each other on an airplane, most other passengers would probably not see much of a connection. But we would know that they both share the "Hero Sidekick Archetype" and are far more related than they might otherwise appear.

As you prepare to read more about the GenoType you've just discovered, you may be wondering how all these odd disparate qualities—tooth shape, jaw angle, head size—could possibly add up to a significant statement about the whole person that is you.

I tell my patients that it's like looking at a bunch of machine parts. Until you know how they fit together, they're just a bunch of odd metal pieces lying in a box. But when they all come together in a well-constructed whole, it's easy to see what they were made for. Just about anybody can recognize a classic Volkswagen Beetle, let alone tell the difference between a Beetle and a Rolls-Royce. But you have to look at the whole, not at the carburetors and transmissions out of context. In the same way, when you put all the different measurements together, suddenly you have *you*, a unique person who nevertheless shares certain basic attributes with hundreds of thousands of people around the globe. So prepare yourself. It's time to meet your GenoType.

# PART III

# The Six Genetic Archetypes

## The GenoType Profiles

# CHAPTER SEVEN

# Meet the GenoTypes

**N**ow you know which GenoType you are, and you may also be thinking about family and friends, wondering which profile fits them best. In a moment, you'll be learning even more about yourself and your loved ones. But first, let me give you a few pointers about how to get the most from reading these profiles.

## Remember That There Are Six Basic GenoTypes— but There Are 7.5 Billion Variations!

As we've seen, the GenoTypes are survival strategies—solutions to problems or reactions to events that our ancestors worked out over the past 100,000 years. They're the result of interactions between genetic heritage, prenatal experience, and our daily interaction with the environment, including diet and exercise. These elements and the ways they interact tend to fall into predictable patterns, which is why I can rely on them to make my suggestions for the six GenoType Diets in Part IV.

Earlier in the book I pointed out that you could look at a collection of individual car parts, but its true "car-ness" isn't visible until all the elements come together, creating the recognizable brands of a Mercedes, or a Porsche, or a Rolls-Royce. But as anyone who drives will tell you, every single car also has its own unique identity, and certainly, once you start driving it, you imprint it with your own particular stamp as well. Your driving style and the environments through which you take your car contribute to the peculiar way its transmission responds to hills or the reliable way its engine always starts on cold mornings.

So be available to the usefulness of these GenoType profiles, but don't stress too much about the details. If something that I say is characteristic of the profile just doesn't fit, then it doesn't. Take it for what it's worth and don't throw out the baby with the bathwater.

## Remember That Every Profile Has Strengths and Weaknesses

As you read through your profile, you're going to find lots of warning signs about the ways things might go wrong for your particular GenoType. Some are prone to cancer, others to heart disease. Some have a tendency to put on excess weight; others struggle with a nervous energy that burns too "hot." Usually, the very qualities that give a GenoType its strengths are also the source of its most disturbing weaknesses. Much like the concept of Yin and Yang in Chinese medicine, each of the strengths of a particular GenoType carries within itself the seed of its weakness, and vice versa.

As a result, some people have a tendency to view their profiles as a kind of death sentence, a blueprint for how things will sooner or later go wrong. They see that their GenoType is prone to cancer or vulnerable to diabetes, and they feel as though I've told them that such diseases are definitely in their future.

Nothing could be further from the truth. To me, these GenoType profiles are like signs along life's highway: "Slow Down—Curves Ahead" or "Slippery When Wet." You want to know what to watch out for so you can take the most effective steps to avoid it. A warning sign on the highway doesn't predict that you *will* have an accident; it only tells you what kind of accidents you should be especially careful to avoid. You don't need the "Curves" sign on a straight stretch of road or the "Slippery When Wet" sign in the desert. You get the warnings you need for the particular dangers you face. And since "nobody gets out of here alive," we all face some!

I'm a naturopathic physician, so my primary interest is always in helping people to achieve the optimal health and vitality that is possible for them. Sometimes it seems ironic that in order to help people make things go right, I have to spend so much time talking about what might go wrong. A long, vital life at your optimal weight is our goal—and I believe you can make great strides in that direction. But in order to move forward, you need a realistic understanding of what might be holding you back. That's what these GenoType profiles are intended to show you— so please view them as keys to your progress, not as predictions of doom.

## There Are No "Combination Types"— Every Type Has Its Own Unique Logic

If you're used to other systems for identifying types—Ayurveda, Chinese medicine, the system of somatotypes developed by William Sheldon— you may be used to thinking that there are broad categories and then many combinations. Ayurveda, for example, has three major types of people—Air, Fire, and Earth—and then four combinations (Air-Fire, Fire-Earth, Air-Earth, and a triple type that partakes of all three).

The GenoTypes don't work like that (though they do overlap with some of the types from other systems). Instead, they represent coherent wholes, six self-sustaining efforts to solve the problems of survival that our ancestors encountered.

I think of the GenoTypes as six different types of tractors, each designed to solve the challenge of a particular type of terrain. One tractor is built very high off the ground so that it rolls right over any stones or stumps it encounters. Its weakness, of course, is that it's not always so stable. Another tractor is built low and squat. You couldn't knock it over if you tried—but when it runs up against even a small stone, it's stuck. There's no one model you can invent that will encounter all difficulties equally well.

Likewise, you don't have an infinite number of solutions. Rather, there's a natural limit to the number of solutions and combinations you could come up with. Once you've given yourself a choice between big, small, and medium wheels, and between wide, narrow, and medium treads, you've pretty much exhausted the possibilities of wheel size—after that, the differences aren't so significant. And since you can't put giant wide-tread wheels on a small, agile tractor that's built for easy steering, or tiny little wheels on a huge, wide, bulldozer-like tractor, you've got a kind of natural limit to the ways the combinations can come out as well.

I'm not saying there's never been a seventh GenoType—or that there may not be an eighth, or ninth as humans continue to live on this earth. But right now, human life as we know it is pretty much defined by these six. How do I know this? Well, when developing the GenoTypes, after running the numbers for the first six, the characteristics simply began to repeat themselves. And just as you can't take the big wheels off your giant tractor and fit them onto your smaller model, so you cannot have combinations among the GenoTypes, even though many of them have common features.

## Six GenoTypes, Three Worldviews

Another thing that's useful to keep in mind as you read through your profile and those of your loved ones is the basic worldviews that define each GenoType. As we've already seen, there are three, each of which defines two of our GenoTypes:

| Reactive Worldview "Inflammation-Based" | Thrifty Worldview "Metabolism-Based" | Tolerant Worldview "Receptor-Based" |
|---|---|---|
| GenoType 1 Hunter | GenoType 2 Gatherer | GenoType 3 Teacher |
| GenoType 4 Explorer | GenoType 5 Warrior | GenoType 6 Nomad |

Let's talk a moment about the definition of "worldview." When most people use that term, they refer to a mental, psychological, or philosophical way of viewing the world, as in "an optimistic worldview." I'm using the word a bit differently—to indicate the specific biological way your GenoType mobilizes itself to respond to the environment. Remember, GenoTypes developed in response to different challenges. Some of the GenoTypes had to hunt for their food; others were able to grow it. Some of the GenoTypes' major challenges were famine and scarcity; others were more concerned with surviving a series of wars. Out of our ancestors' attempts to meet the challenges of their time, our GenoTypes developed—and our worldview along with them.

As we've seen, a *reactive* worldview responds to the environment in an aggressive, proactive, and even hostile way. Kill that game! Destroy those invading microbes! Run through the forest and find that prey—otherwise, we'll go hungry tonight! Muscles, bone length, heart and blood pressure, and immune system all work together to enable reactive Geno-types—Hunters and Explorers—to make the most of this approach. This worldview is *inflammation*-based. The level of reactivity found in Geno-Types with this worldview is programmed in early childhood. Was your childhood full of antibiotics? Did you grow up in the city? Were you an only child? Were you formula-fed rather than breast-fed? If you can answer yes to these, you might be one of the reactive GenoTypes. The downside of this worldview is that it comes at the expense of damaging the body's own tissues as a by-product of all this reactivity, the type of friendly fire that leads to autoimmune disease.

A *thrifty* worldview responds in a cautious way, seeking not to confront threats but to avoid them. Conserve those calories, says this approach. You don't know when your next meal is coming, so avoid exertion when you can. Life is full of calamities, so keep an eye out for the best route to self-preservation. A thrifty worldview is highly desirable in a world of shortage or scarcity, but it is poorly adapted to the luxuries of today's widely available fats and sugars. Thrifty GenoTypes—Gatherers and Warriors—have a *metabolic* basis to survival. They respond to scarcity by slowing down their metabolisms—in particular, the way in which the cells respond to hormone stimulation. Thrifty GenoTypes often have normal levels of hormones yet show all the signs of hormone insufficiency. The *outside* of the cells get the proper hormone stimulation, but the *inside* of the cell can't respond to it.

A *tolerant* worldview is accepting and adaptable, geared to people who had to travel through different environments and confront an ever-changing world. What worked yesterday may not work tomorrow, so don't react too quickly—think it over, figure it out, says this approach. This worldview is constantly adapting its responses to the environment, often by altering the binding sites, or *receptors,* that are found on the cells and tissues. Many of these receptors are used by microbes to attach to the tissues and organs, sometimes for good (such as our relationship with the "good bacteria," or probiotics, in our digestive tract), other times not so good (such as when we catch a cold or pick up a parasite). If you put up your guard against every new microbe or bacterium, there are a lot of new foods you won't be able to eat and a lot of new places that will make you sick, so try to get along with whatever the environment throws at you, adapting rather than defending yourself.

Now, as you can see, I've used language that can describe both physical attributes and mental and emotional qualities. To some extent, that's because there *is* a correspondence between our physical attributes and our other qualities, if only because our minds are also physical. Every emotion has a physical expression in neurons and hormones and biochemicals. Every biological response—stress, hunger, the watering of our eyes—has its emotional correlate. Feeling happy may cause you to smile, but it's also true that if you smile, you start to feel happier. Anxiety may make your heart race, but if coffee produces the same physical effect, you will probably also feel anxious, perhaps without knowing why. Our bodies, minds, and emotions all live very close together.

Accordingly, our "reactive," "thrifty," and "tolerant" selves may have developed as physical responses to the planet's challenges, but they have come to express something about our mental and emotional responses as well. Perhaps, too, they were the psychological characteristics that were most helpful for survival in various situations and so were passed along from one generation to the next.

Get ready to learn more in Chapters 8–13, where you'll find out everything you need to know about the GenoTypes' strengths, weaknesses, metabolism, and health concerns, as well as each GenoType's specific challenges when it comes to achieving and maintaining an optimal weight.

# CHAPTER EIGHT

# GenoType 1:
# The Hunter

**T**all, thin, and intense, with an overabundance of adrenaline and a fierce, nervous energy that winds down with age, the Hunter was originally the success story of the human species. Vulnerable to systemic burnout when overstressed, the Hunter's modern challenge is to conserve energy for the long haul.

| Typical Features of the Hunter | | |
|---|---|---|
| Psychological | Biometric | Biochemical |
| • Mechanically inclined, detail-oriented, heightened sense of fair play<br>• Adrenaline-driven<br>• Terrific at handling stress when properly nourished and in balance—but when poor diet and stress overload kick in, watch out! The Hunter's pituitary-adrenal glands go into overload and even a little stress becomes too much. | • Symmetrical—both sides of the body seem to be the same<br>• Often has white lines on fingerprints, indicating digestive problems<br>• Tend to be ectomorphic or meso-ectomorphic<br>• Ring fingers tend to be longer than index fingers<br>• Front teeth tend to be shovel-shaped<br>• Squarish jaw<br>• "Andric"—tends toward masculine body type | • Always blood type O<br>• Reacts intensely to PROP test strips ("super-tasters") |
| Superstar Hunters | • Thomas Jefferson (U.S. president)<br>• Katherine Hepburn (actress)<br>• Diana (Greek goddess of the hunt)<br>• Michael Jordan (basketball player) | |
| Slogan | "Shoot first, ask questions later." | |
| Strengths to Count On | • High energy levels<br>• Metabolically efficient—when in good health can eat heartily without gaining weight and will direct calories just where they're needed for optimal strength, stamina, and well-being<br>• With the right diet, a tremendous cellular powerhouse whose body functions at peak efficiency<br>• Athletic; tall, strong-boned, and well-formed | |

| Typical Features of the Hunter *(continued)* | |
|---|---|
| **Weaknesses to Watch Out For** | • Hair-trigger response to infections, viruses, allergens—tends toward autoimmune responses<br>• Without the right diet or stress relievers, vulnerable to midlife burnout and a tendency to age ungracefully<br>• May have digestive problems and poor assimilation of nutrients<br>• Stress organs are especially vulnerable: adrenals, pituitary |
| **Health Risks** | • Allergies<br>• Autoimmune conditions, such as asthma or rheumatoid arthritis<br>• Depression<br>• Joint problems<br>• Celiac-like gut problems<br>• Reproductive cancers (more in men) with advancing age |

When I first met Matt, a modern-day GT1 Hunter, he could barely sit still on my examining table. He drummed his fingers nervously on his knees and his legs were twitching—never-ending legs that seemed not to be able to stop moving. Naturally long and lean, Matt looked almost emaciated, his face haggard, blue-black shadows smudged beneath his eyes. He hadn't been sleeping well, he told me—when he was stressed, as he had been lately, sleep was always the first thing to go, except maybe appetite. He hadn't had much appetite either, and what he had eaten he had trouble digesting, his gut recoiling at the first hint of food. As a result, Matt suffered from irritable bowel syndrome and contended with bouts of diarrhea so severe that his medical doctor had seen no alternative but to give him sedatives in a futile effort to get his digestive tract to relax.

As we spoke, I felt a growing sense of Matt's uneasiness, a never-ending sense of anxiety that his frequent sessions at the gym had seemed to make worse, not better. "I've been like this since I was a kid," he confessed. "Sometimes I get so nervous it feels like my heart wants to jump out of my throat." When I took hold of his hand to take his pulse, I felt how dry and scaly his skin was and was not surprised to find that his fingerprints were shot through with white lines. White lines in the fingerprints are a sure sign of gluten intolerance, but Matt assured me that, too, was a normal condition to which he'd never paid much attention.

It was clear to me from Matt's symptoms that here was a classic Hunter gone wrong, his natural strength, agility, and high-powered energy turned against itself. Instead of taking advantage of the Hunter's marvelous capacity to respond creatively to stress, Matt had converted his stress into more stress; instead of using his energy to solve new problems, Matt converted it all into anxiety.

His family history only confirmed my diagnosis, since many in his family also suffered from classic Hunter disorders. His mother had rheumatoid arthritis, his father had low thyroid function, and two of his three sisters were prone to depression.

The bad news was that Matt was being slowly done in by the worst aspects of his GenoType. The good news was that there was still plenty of time for him to turn the situation around and reclaim his Hunter strengths. I put Matt on the GT1 Hunter diet—the same one you'll find in Chapter 15—and told him to take heart. Within a few weeks, I assured him, most of his symptoms would be vastly reduced, if not altogether gone, and he'd soon be feeling better than he'd ever imagined.

Despite my optimism, even I was surprised by the Matt who walked into my office three months later. Gone was the haggard face and shadowed eyes: Now Matt's eyes sparkled with new vitality, and his face had the healthy glow of someone who's begun to regenerate the tissue beneath the skin. Gone were the digestive problems: Matt now ate happily and with good appetite. Gone, too, were problems that Matt hadn't even realized he'd had. In his late forties, Matt had come to view stiff joints and a certain amount of joint pain as the inevitable price of aging, but now, he told me, he felt limber and flexible and his joints never bothered him.

Matt felt calmer, too, and far better able to handle whatever life threw at him. When he'd first come to see me, Matt had been one of those people who have trouble handling even *good* stress—the excitement of a long-desired vacation, the chance to work with a highly valued colleague, an effort by his wife to rekindle their twenty-year romance. Now, though,

Matt seemed to have regained the Hunter's natural agility in responding to a wide range of stimuli, that basketball player's talent for maintaining a 360-degree circle of awareness. Whether encountering "good" stress or "bad," Matt was newly able to respond calmly and effectively, discovering within himself a strength and effectiveness he'd never imagined could be his. He found, too, that his long sessions at the gym now made him feel better—energized and healthy—instead of exhausted and depleted, as they had before.

No longer suffering from the Hunter's typical weaknesses, Matt had managed to claim the greatest strengths of his GenoType. I couldn't have been happier for him—or for you Hunters who are reading this book, ready to claim the power of your own GenoType.

## Hunters at Their Best

If you recall that two Hunter Superstars are Michael Jordan and Thomas Jefferson, you'll quickly grasp the extraordinary strengths of the healthy Hunter. This GenoType has a great capacity for responding to stress in creative, effective, and agile ways. The perfectly tuned Hunter has a tremendous amount of energy; quick, sharp powers of perception; and a remarkable ability to adapt to rapidly changing situations. If you imagine the total awareness of the basketball player—the 360-degree awareness of events, the rapid response to each new play, the hair-trigger leaps and dashes and passes—you'll see that the Hunter is primed for physical activity and geared to *react*. Like any great athlete—or, for that matter, like a filmmaker, a firefighter, or an emergency rescue worker—Hunters are good at short, sustained bursts of activity as they respond to a series of crucial, rapid, and unexpected demands that require their absolute best. Of all six GenoTypes, Hunters are best suited to living in the present, springing instantly into action like a cat in front of a mouse hole—or, perhaps, a lioness in pursuit of tonight's dinner.

The Hunter is also a great detail person, with the kind of mind that can break every object into its component parts, every activity into a series of small steps. Here's where the Thomas Jefferson aspect of the

Hunter comes in—the mechanical, detail-oriented mind that could lay out in concrete terms what independence meant for a new nation and how thirteen separate states would have to learn to work together.

## Problem Areas for the Hunter

As with all GenoTypes, Hunters' greatest strengths are also their most vulnerable weaknesses. Michael Jordan is a terrific player in an intense two-hour game—but I'd hate to see what might happen to him stuffed in a cubicle for ten hours a day working for a demanding boss or after a grueling three-month research project. I'm not saying he couldn't handle it, but he'd have to take extra care not to burn out from the long, sustained periods of concentration and the lack of physical release.

Likewise, if you put Jordan on the court after spending two hours in bumper to bumper traffic before the game, he might not respond with the same agility and grace. Unless he were very careful about stress release, diet, and how he used his body, he might find all that physical brilliance converted into nervous energy, an engine that burns itself out rather than efficiently burning fuel.

The out-of-balance Hunter's detail-oriented gifts can turn into a somewhat obsessive focus on details, as with the manager who gets bogged down in office regulations and loses the big picture of what the company is meant to accomplish. By the same token, the present-oriented Hunter can become derailed by anxiety when forced to confront the pain of the past or the uncertainty of the future.

Hunters have been optimized for long periods of nothing happening punctuated by short periods of intense stress—think of that cat crouched before a mouse hole. They don't do so well with modern life's tendency to hand us long periods of low-level stress that never end. That's why Hunters have to take extra care to release stress physically while never falling into the trap of excessive exercise. Likewise, they must learn how to cope with stress out of a strength that insists on its own terms—the cat that waits for the right moment to pounce, rather than the one that frantically bats its paw at the mouse hole, hoping to force the mouse to come out.

The organs that govern our stress response are the adrenal and pituitary glands, and, not surprisingly, these are vulnerable areas for most Hunters. Well-functioning Hunters run on a healthy adrenaline high, with short, sustained bursts of energy that members of other GenoTypes often find astonishing. But good adrenal function requires downtime—periods when adrenaline is discharged through satisfying physical exertion and when the mind returns to a place of calm. Hunters in our modern world, with its perpetual deadlines and sedentary life, are all too prone to adrenal burnout, the sad condition that results from excess adrenaline production and insufficient stress release. At that point, Hunters find it extremely difficult to mobilize their energies, and instead of healing, vigorous exercise only wears them out further. That's why it's so important to follow the GT1 Hunter diet, which will nourish your adrenal and pituitary glands, and why you need to follow the recommendations for exercise and lifestyle as well.

## The Hunter Metabolic Profile

At their best, Hunters have an absolutely superb metabolism—perhaps the best of the six GenoTypes. They have a positive genius for converting calories into the perfect combination of muscle, bone, and fat, and their physiques are primed for optimal use of their lean, athletic limbs and long, strong backs. If you're a Hunter, you might think of yourself as a top-of-the-line sports car that burns high-octane fuel—and then ask yourself what happens to that Porsche if it gets poor-quality gasoline or isn't driven at the top speeds it was designed for.

Hunters are typically "andric," meaning that they tend toward having been stimulated in utero by the male androgens. This, in combination with their abundant levels of growth hormone in early childhood, tends to give them a long, rangy look. Hunters typically have legs that are longer than their torsos, and lower legs that are longer than their uppers.

The GT1 Hunter diet is designed to help Hunters burn high-quality calories while calming and soothing their hyperreactive digestive and immune systems (about which more in a moment). So just like that high-end sports car, which lasts a lot longer if you give it high-octane gas, Hunters will see enormous benefits in how they age if they eat right and get the types of exercise that release stress rather than wear them out.

## The Hunter Immune System Profile

The Hunter is one of the great success stories of human evolution, with an immune system designed to mobilize the body's entire force against any toxic invader. In the era before antibiotics, it paid to have an immune system that brought out the big guns against bacteria, not to mention viruses, allergens, and any other threats. For most of human history, the Hunter's motto of "Shoot first, ask questions later" served this GenoType extremely well.

The downside, obviously, is that all that shooting takes its toll. A hyperreactive immune system is a blessing when the world actually is full of toxic invaders; in a world of dust, cat dander, pollen, and yeast, it's not so useful. Hunters are prone to inflammation—the heat, redness, swelling, and pain that result when the body fights off what it perceives as a dangerous invader. In many cases, the cure is worse than the disease, since inflammation contributes to numerous health problems, including arthritis, asthma, cancer, diabetes, and heart disease. Inflammation may also contribute to obesity, so hyperreactive Hunters' price for not following their ideal diet is to, well, need a diet!

As you can see, an overreactive immune system is the powerful Hunter's Achilles' heel, so the GT1 Hunter diet is designed to get that immune system back in balance and damp down its hair-trigger responses when they're not really needed. (My wise Hunter wife calls it being *responsive* rather than *reactive*.) Interestingly, the sturdy Hunter is *not* particularly sensitive to environmental chemicals (that's the bailiwick of GT4 Explorers). The degree of reactivity seen in each individual Hunter is a combination of many pre- and postnatal influences. Did you grow up in an urban environment? Were you an only child? Take lots of antibiotics? Did your mom feed you formula from a bottle rather than breast-feed? Answering yes to any of these will point a Hunter toward greater reactivity.

## The Hunter GenoType Diet

The Hunter is blessed with a metabolism that makes it easy to lose weight and to maintain an ideal weight. So if you are overweight, thanks to either poor habits or to the inflammatory conditions described above, you'll find it easy to correct that problem once you adopt the Hunter GenoType diet.

Hunters have a lot of natural resilience, which the Hunter diet will help you activate. If you haven't been eating right for your GenoType, you'll probably have done some damage to the lining of your digestive tract—Matt certainly had—and having worn fingerprints with lots of white lines will let you know that's the case. But again, you've got lots of room to reverse the situation if you eat the foods that will restore your gut's natural protective mechanisms—and the GenoType 1 Hunter diet will help you do just that.

## Hunter Diet Dos and Don'ts

The GT1 Hunter diet is a carnivorous, low-lectin, low-gluten diet. ***Hunter Dos*** are superfoods and supplements that epigenetically heal their digestive tracts, help them to handle stress better, and control runaway inflammation.

### Hunter Diet Dos

The best superfoods for the Hunter GenoType have nutrients in them that:

- **Have the nutritional building blocks needed for genetic improvement.** These superfoods are rich in purines and nucleotides from protein and cultured foods.
- **Increase muscle mass and decrease body fat.** The prime foods for hunter weight loss are identified by a diamond (◊) icon.

- **Heal and regenerate the digestive tract.** These superfoods are rich in butyrate, a fatty acid known to exert a nutritive effect on the digestive tract. Butyrate also has highly desirable effects on gene function.
- **Dampen inflammation.** These superfoods include "clearing foods," which lower reactivity to allergens and lectins in the diet.
- **Enhance stress-handling abilities.** These include sterols found in plant foods and amino acids such as tyrosine found in meats. These nutrients control stress at both the physical and emotional levels.
- **Are rich sources of tissue-protecting antioxidants.** These phytochemicals mop up tissue-damaging free radicals and slow down the Hunter tendency toward rapid aging.

## Hunter Diet Don'ts

*Hunter Don'ts* are foods that are best minimized or outright avoided. The *Hunter Don'ts* exclude from the diet those foods that:

- **Slow down the Hunter metabolism.** Many grains, nuts, and seeds can interfere with the proper function of insulin, causing even the normally lean Hunters to have difficulty keeping their weight down.
- **Are high in the "bad" fats.** Bad fats increase inflammation and can cause damage to the walls of the arteries. Trans fats are bad for the arteries; an undesirable ratio of omega-6 to omega-3 fats can increase the levels of inflammation.
- **Are too high in simple sugars.** High levels of sugars encourage bacterial overgrowth, which increases inflammation in the digestive tract.
- **Irritate the gut.** Many foods contain ingredients that can irritate the lining of the Hunter's gut, causing fatigue and inflammation. Many mold-containing foods and fungi can cause increased inflammation in Hunters.
- **Contain gluten, a lectin, or another allergen:** Gluten is a protein found in many grains that can irritate the gut lining in sensitive individuals. Chitinase is an enzyme that can initiate allergic reaction in the gut. It is found is some nuts and fruits. Lectins are proteins that can

interfere with proper digestive and immune function. Phenols are plant compounds that cause allergic reactions in many Hunters.

Some foods on the *Hunter Don'ts* list need only be avoided for a short period of time so that the Hunters can regain their balance. After 3–6 months, you can reintroduce them back into your diet in modest amounts. These foods are identified by a black dot (•) icon. If you are battling an illness or coming down with something, you may want to ramp up your compliance by avoiding these foods for a while.

## Unlisted Foods

Unlisted foods are foods that don't appear to do much good or bad. They are essentially neutral, and can be used judiciously (2–5 times weekly). The nutrients in them will benefit you, but they won't specifically help you restore balance to your genes or health to your cells. Feel free to eat them—but don't neglect the foods I recommend. The GenoType Diet is always evolving and I'm frequently adding new foods, so check in with my Web site (www.genotypediet.com), especially if you have questions about a specific food.

Well, Hunter, now is the time to put word and intent into action. From here you'll skip to Chapter 14 to learn how you can get the most from the GenoType Diets. After that, we're off to Chapter 15 and the Hunter food, supplement, and exercise prescriptions.

# CHAPTER NINE

# GenoType 2: The Gatherer

Gatherers carried humanity on their backs during times of famine and scarcity. They are Nature's ultimate survival strategy. Vulnerable to conserving calories as stored fat, the Gatherers' modern challenge is to fit their survival programming to the realities of today's overabundance of fats and sugar.

| Typical Features of the Gatherer | | |
|---|---|---|
| Psychological | Biometric | Biochemical |
| • Phenomenal capacity for prolonged and concentrated brain work<br>• "Algorithmic" mind-set: natural-born problem-solver<br>• "Early adopter" of new and revolutionary ideas<br>• Sweet-natured with a tendency toward emotional "highs and lows"<br>• "Exercise-challenged" | • Endomorphic body type: always looks "padded," even when at proper weight<br>• Tends toward high BMI and waist-to-hip ratio<br>• Lower leg is shorter than upper leg<br>• Longer index fingers than ring fingers<br>• "Gynic"—narrow interior space between legs<br>• Asymmetrical finger-print patterns—one hand not matching the other<br>• Often has extra cusp on first molar<br>• Almond-shaped jaw | • Blood type O or B<br>• Mostly Rh-positive<br>• Almost always PROP "non-tasters"<br>• Often "non-secretors"<br>• High estrogen levels |
| Superstar Gatherers | • Oprah Winfrey (media personality)<br>• Orson Welles (film director)<br>• Marilyn Monroe (actress)<br>• Elvis Presley (singer, actor) | |
| Slogan | "Whoever dies with the most wins." | |
| Strengths to Count On | • Terrific mental endurance<br>• Highly motivated<br>• Fertility—male and female<br>• Potential to age well | |

| Typical Features of the Gatherer *(continued)* | |
|---|---|
| **Weaknesses to Watch Out For** | • Gatherers are unsuccessful crash dieters with a strong tendency to store calories as fat.<br>• Appetite regulation can be a problem.<br>• Elevated estrogen sensitivity can stimulate hormonal cancers.<br>• Accumulation of damaging chemicals in tissue, which can lead to diabetes, high blood pressure, and Alzheimer's disease. |
| **Health Risks** | • Alzheimer's disease<br>• Depression<br>• High blood pressure<br>• Insulin resistance and diabetes<br>• Low thyroid activity<br>• Obesity<br>• Reproductive cancers (more in women) with advanced age |

One look at Carmen told me she was special. Born in Puerto Rico, she literally pulled herself up by her own bootstraps: Her mother had left school at age twelve, given birth to Carmen at sixteen, and was dead from a heroin overdose at age nineteen. Raised by loving grandparents in New York City, Carmen started out cleaning hotel rooms, but finished school at night and now ran a large hospital cafeteria. A devout Jehovah's Witness, Carmen was coming to the clinic to seek our treatment for high blood pressure and to discuss some recent disturbing blood sugar readings. Short and rather stocky, Carmen had admitted to trying just about every diet out there. In all cases, she would lose weight for a while, but in time it would all come back. In addition, these crash diets seemed to make her look worse than before, often causing alarm in her children.

Carmen could provide no information on her father's health history and knew only from her grandparents that her mother's pregnancy was quite difficult: The mother had had a lot of nausea and could not keep much food down. Carmen suspected that she had smoked during the pregnancy. Her grandmother had died four years earlier of advanced Alzheimer's disease. Carmen had done a much better job raising her own kids. One, the boy, had become an accountant for a large corporation. The girl, also a college graduate, had just given birth to a beautiful and health baby boy.

A quick look at Carmen showed all the classic signs of a GT2 Gatherer. Comparing her leg lengths, we could easily see that her upper legs were longer than the lower—often a sign of diminished nutrition during fetal growth. Carmen also had that "padded" look you often see in Gatherers: The skin in certain parts of the body, like the wrists and backs of the hands, hides the underlying tendons.

When we sat down in my office for the consultation, I explained to Carmen the nature of her "thrifty" metabolism and how her mother's difficulties during the pregnancy might now be "coming home to roost" and how crash dieting would most likely make her sicker, not healthier. "Oh, Doctor," she exclaimed, "I have been praying for someone to tell me what to do."

You could not ask for a better patient than Carmen. We would meet every three months, and at each meeting she had better glucose readings, lower blood pressure, and a consistent loss of weight. Moreover, Carmen *looked great!* On her last visit before I discharged her, she showed me a picture of herself at age twenty-two. If I ignored the different hairstyles and the obvious changes in fashion, I would not have been able to distinguish between the two.

## Gatherers at Their Best

As a GT2 Gatherer, your strength is centered on the concept of acceptance. Although it is possible that you're a reed-thin fashion model and saunter down a catwalk, your basic genetic makeup makes it unlikely. You wouldn't feel well at that degree of thinness, and you wouldn't look that well, either. You have a different mission in life than simply to lose all the weight you can as fast as you can. Weight loss can happen under carefully controlled circumstances, but it can't be your only goal.

Because you have such a determined and dogged nature, you have great potential for genomic improvement. If you nourish yourself properly, you can temper the thriftiness trait in yourself and even pass the improvement on to your children and grandchildren. In effect, the truly accomplished GT2 Gatherer announces, "The thriftiness gene will be altered here!"

Gatherers often have wonderful, warm, sensuous personalities. An affinity for home, hearth, people, and food makes them beloved friends and partners. They have nurturing and forgiving natures. They are intellectually adventurous and are often among the first to adapt to new ideas and methods. Although easygoing, they are highly principled and often tenacious defenders of justice.

The combined influences of nature and nurture give Gatherers a complex psychological profile. Let's face it, our attitudes are formed in large part by social feedback, and in Western cultures there has been a reversal of fortune for Gatherers. One hundred years ago, even industrialized societies revered voluptuous women and portly men as symbols of affluence and fertility. Today the opposite is true. Being slender is viewed as a sign of success and affluence.

## Problem Areas for the Gatherer

The Gatherer's metabolic thriftiness, such a survival asset in ancient times of scarcity, comes with a price to pay. It doesn't fit the lifestyle of the modern industrialized world, with its amply and inexpensively available carbohydrates and fats. The vast majority of these excess carbohydrates and fats will be taken out of the bloodstream and stored. However, Gatherers are usually so good at taking sugar out of the bloodstream and storing it that they essentially spend most of their time in a permanent state of hypoglycemia. They suffer on two accounts: First, the energy sources are stored instead of burnt, resulting in weight gain; and second, they don't get the "reward" of having consumed these nutrients because they are removed so efficiently from the bloodstream that the brain and muscle tissue fails to get their fair share.

This eventually fractures the relationship between food and appetite to the point that Gatherers begin to eat simply to feel better rather than as any result of hunger signaling. The effects are chronic and serious. Gatherers can easily develop disturbances of their carbohydrate regulation and insulin sensitivity, resulting in metabolic Syndrome X and "diabesity." It is a slippery slope from there to artery disease, kidney disease, and premature aging.

Gatherers usually manifest the physical signs of thriftiness. They are not very tall, and their lower legs are shorter than their upper legs. Thrifty genes tend to inhibit the activity of insulin-like growth factors, both in utero and in early childhood. Growth factors are molecules that are involved in many key aspects of development. Among their many functions, these growth factors cause the elongation of lower leg bones, and Gatherers' typically shorter lower legs are a sign of their inhibition.

Asymmetry is another characteristic of the Gatherer. These often show up by one hand having different fingerprint and palm patterns than the other. In particular, Gatherer women tend to have visible differences in the size of their breasts.

A common health challenge seen in Gatherers is hypothyroidism, or low thyroid hormone production. GT2 Gatherers tend to be PROP and PTC non-tasters, indicating underactivity of the thyroid gland, whose job it is to control the metabolic processes. Hypothyroidism causes fluid retention, muscle weakness, and low body temperature.

## The Gatherer Metabolism

An endomorph can easily become overweight, but being an endomorph in itself is not the same as being overweight. Marilyn Monroe was probably an endomorph, and I've never heard anyone refer to her as fat, but rather as lush and voluptuous.

If their metabolic thriftiness is corrected through diet and proper lifestyle, Gatherers can age very well. However, there is a direct link between diet and aging. Gatherers who let their thriftiness go unchecked, who rob their systems of calories with extreme weight-loss diets, and who continue to accumulate fat stores are going to do long-term damage to their bodies. The surfaces of their cells literally become gummed up with sugar and fat or sugar and protein complexes, preventing them from performing properly.

Gatherers are typically "gynic," exhibiting tendencies associated with above-average estrogenic stimulation in the womb. Gatherers almost always have long index fingers compared to ring fingers, indicating high levels of estrogen in the womb. Other signs of estrogenic influence are the narrow interior space between the legs, and a wider, rounder jaw angle. These attributes must have made female Gatherers highly desirable, especially when combined with their low waist-to-hip ratio (which imparts an hourglass shape to the figure). Not to be undone, male Gatherers also often impart a striking presence. Rent the movie *Jailhouse Rock* with Elvis Presley, watch his performance of the title song, and you'll see what I mean.

## The Gatherer Immune Profile

Gatherers typically have strong immune systems. However, there are specific weaknesses that should be addressed. Because of the extra-estrogenic activity during fetal development, there are slightly greater odds of problems with the reproductive organs, especially estrogen-dependent cancers. When they do strike, reproductive cancers tend to afflict Gatherer women earlier in life.

Because Gatherers are so good at storing fat, they can be more at risk of accumulating man-made chemicals called xenobiotics—a term that literally means "foreign to life." Virtually all man-made chemicals are xenobiotic. Principal xenobiotics include drugs, carcinogens, and various compounds that have been introduced into the environment by artificial means, such as pesticides, fertilizers, and hydrocarbons. Periodic detoxification is a great way to prevent these complications from developing.

## The Gatherer GenoType Diet

We've all seen Gatherers on weight-loss programs. Just think of any person you knew both before and after a period of extreme weight loss. Those who looked the worst and felt the most miserable after dropping all of that weight were Gatherers, especially those who developed puffy, darkened bags under their eyes as a bonus to go with their newly svelte figures. In reality, the best thing Gatherers can do for their dietary health is to eat enough food. However, while Gatherers need to consume enough food, it also has to be the right kind of food. Like Carmen, they must learn not simply how to cut calories but also which specific foods do the healing—and the Gatherer GenoType Diet will help you do just that.

## Gatherer Diet Dos and Don'ts

The Gatherer GenoType Diet is a high-protein, low-glycemic diet. **Gatherer Dos** are superfoods and supplements that epigenetically reprogram their thrifty genes, help cleanse the cells of accumulated metabolic byproducts, and restore the sensitivity of their cells to the body's hormones.

### Gatherer Diet Dos

The best superfoods for the Gatherer GenoType have nutrients in them that:

- **Have the nutritional building blocks needed for genetic improvement.** These superfoods are rich in the vitamins, minerals, purines, and nucleotides from proteins and cultured foods.
- **Increase muscle mass and decrease body fat.** This improves the metabolic rate, which speeds up proper weight loss. The prime foods for Gatherer weight loss are identified by a diamond (◊) icon.
- **Cleanse the fat tissue of undesirable toxins.** Gatherers can accumulate man-made (xenobiotic) toxins. We want to emphasize foods that help eliminate them.

- **Enhance sensitivity to their metabolic-boosting hormones.** This increases your metabolic rate, which speeds up proper weight loss. The prime foods for Gatherer weight loss are identified by a diamond (◊) icon.
- **Remove accumulated metabolic by-products from the cells.** These foods contain ingredients that speed the removal of advanced glycation end products (AGE), the "burnt sugar" molecules that accumulate so easily in Gatherers as they grow old.

## Gatherer Diet Don'ts

*Gatherer Don'ts* are foods that are best minimized or outright avoided. *The Gatherer Don'ts* exclude from the diet those foods that:

- **Slow down the Gatherer metabolism.** Many grains, nuts, and seeds can interfere with the proper function of insulin, causing the Gatherer to lose weight with only the greatest difficulty.
- **Enhance the deposition of cellular debris.** Foods known to enhance the production of AGEs are excluded from the Gatherer diet.
- **Are high-glycemic foods.** High-glycemic foods produce large fluctuations in blood glucose and insulin levels. Avoiding these is the secret to reducing your risk of heart disease and diabetes and is the key to sustainable weight loss.
- **Block proper hormone stimulation.** Gatherers rarely require hormone therapy. More often, they require that something be done to restore sensitivity to their own hormones. Foods that interfere with optimal hormone function in the Gatherer need to be avoided, especially in the early stages of the diet.

Some foods on the *Gatherer Don'ts* list need only be avoided for a short period of time so that you can regain your balance. After 3–6 months, you can reintroduce them back into your diet in modest amounts. These foods are identified by a black dot (•) icon. Of course, if you are battling an illness or feel your weight beginning to creep back up, you may want to ramp up your compliance by avoiding these foods again for a while.

## Unlisted Foods

Unlisted foods are foods that don't appear to do much good or bad. They are essentially neutral, and can be used judiciously (2–5 times weekly). The nutrients in them will benefit you, but they won't specifically help you restore balance to your genes or health to your cells. Feel free to eat them—but don't neglect the foods I recommend. The GenoType Diet is always evolving and I'm frequently adding new foods, so check in with my Web site (www.genotypediet.com), especially if you have questions about a specific food.

Well, Gatherer, now is the time to put word and intent into action. From here you'll skip to Chapter 14 to learn how you can get the most from the GenoType Diets. After that, we're off to Chapter 16 and the Gatherer food, supplement, and exercise prescriptions.

# CHAPTER TEN

# GenoType 3: The Teacher

**S**inewy and flexible, with an amazing adaptability, the Teacher is a balance between opposing and often contradictory forces. Blessed with a tolerant immune system, the Teacher can be burdened by excess altruism, leading to problems finding and dealing with the bad guys.

| Typical Features of the Teacher | | |
| --- | --- | --- |
| Psychological | Biometric | Biochemical |
| • Natural exuberance and a calm, steady way of looking at the world<br>• Soul of an artist<br>• Deep relationship with nature<br>• Meta-analytical—can "see the forest for the trees" | • "Sinewy"—can usually see tendons under the skin<br>• Torso usually longer than legs<br>• Generally of moderate to short stature<br>• Index-to-ring-finger ratio often inverted from hand to hand<br>• Squarish jaw angle<br>• High number of whorl-type fingerprints<br>• Extra molar cusp common<br>• "Andric"—tends toward masculine body type | • Often blood type A, occasionally blood type AB<br>• Secretor<br>• Rh-positive<br>• PROP/PTC "taster" or "super-taster"<br>• Tolerant immune system<br>• Tend toward excess bacterial overgrowth in the digestive tract |
| Superstar Teachers | • Abraham Lincoln (U.S. president)<br>• Athena (Greek goddess of wisdom)<br>• Morihei Ueshiba (founder of the Japanese martial art of Aikido)<br>• Che Guevara (revolutionary) | |
| Slogan | "Why can't we all just get along?" | |
| Strengths to Count On | • Successful environmental adapter<br>• Powerful spiritual energy<br>• Ages gracefully<br>• Tremendous mechanical strength coupled with flexibility | |
| Weaknesses to Watch Out For | • Sensitive digestive system<br>• Immune system sometimes doesn't catch cancer mutations in the early stages<br>• Tends to tolerate bad microbes rather than eliminate them<br>• Can become excessively detail-oriented | |

| Typical Features of the Teacher *(continued)* | |
|---|---|
| **Health Risks** | • Chronic childhood ear infections<br>• Chronic lung, stomach, and bowel problems<br>• Bacterial infections<br>• Potentially high risk for breast cancer in later life |

"You can call me Harry" came back in response to my ineffectual attempts to pronounce Haruo's full name. "My first name means 'springtime man' in Japanese, and that's why I'm here. I think I need some spring cleaning."

Harry's face literally lit up as innumerable gold teeth flashed an infectious lopsided grin, and I was treated to the fascinating story of his life. He was Japanese but had grown up in China, where his parents were employed as railroad technicians during the Second World War. Both had died, as had two of his grandparents, from cancer of the stomach. After the war, he immigrated to the United States, where he was employed for many years by a large American manufacturer of appliances. His wife had recently passed away from cancer, and he was here at the clinic at the behest of his daughter, who was studying to be a naturopathic physician at one of the colleges in the Pacific Northwest.

Harry had always taken great pride in his appearance, to the point of having even his shoes and hats custom made. Recently, he had noticed a rather odd fact. His hats were no longer fitting his head. "Perhaps I'm getting smarter," he joked. Physical examination showed nothing special, other than the fact that the guy was in great shape for his age. "I love to garden—perhaps too much. I keep almost two full acres under cultivation. All done by hand. But I'm getting more tired these days. " Harry was blood type A and had one other interesting lab finding. The amount of hydrogen in his breath was astoundingly high. High levels of hydrogen in the breath almost always signifies a bacterial overgrowth in the digestive tract. Most of us have little or none. Harry's was the highest I'd ever recorded.

Concerned about Harry's expanding head, I sent him to the radiologist for X-rays. Sadly, the findings confirmed my suspicions. Harry had a cancer called multiple myeloma and it was altering the bone structure of his skull. He took the news surprisingly well. "I've lived a good life. Now, Doc, what can we do?"

I explained to Harry that although the disease had a high mortality rate, a small percentage of patients do rather well and live quite a while with the illness. "Good. I'd like to be in that group. Someone has to be in there, right, Doc?"

After our measurements and analysis were complete, it was obvious that Harry was a Teacher. Almost all Teachers have a tolerant immune system and lots of bacterial overgrowth. I explained that the approach would be supportive and that we were going to try to put some of those cancer genes to sleep while we reactivated some of the watchdog genes that are supposed to keep an eye out for this sort of thing.

Three years have passed and Harry still tends his garden. He's also met a nice woman and remarried. When one of the clinic residents asked him why he had done so well, he just smiled his golden smile and replied, "I just do what the Doc tells me to do and take every day one at a time."

## Teachers at Their Best

In peak condition, Teachers have a natural exuberance and a calm, steady way of looking at the world. Teachers are first-class adapters. Long ago, Teachers easily incorporated rich and abundant sources of protein in combination with rapid adoption of simple but effective agricultural technologies, and they went on to develop a reasonably flexible worldview. Their biological tolerance is reflected in their personalities. They have a centered, calm manner. In general, Teachers have the souls of an artist and are happy and healthy as long as there are ample avenues for creative expression in their lives. It is no surprise to me that many of my Teacher GenoType patients have a deep involvement with meditation, tai chi, or yoga. Teachers have a special life force or spiritual energy, known as *Chi*, and a deep relationship with nature. In effect, tolerance becomes a form of coexistence.

Teachers are great at meta-analysis, the ability to evaluate numerous types of data and synthesize its essence, or gestalt. Because it is such a tolerant GenoType, Teachers are not all that inflammation-prone and in general seem to be less allergic.

Teachers tend to age well, and many reach significantly advanced years. But this doesn't happen automatically. The key for Teachers is to seek balance in everything—whether it be diet, work schedules, sleep-wake cycles, or methods of exercise.

With the right diet and lifestyle changes, the results of following the GT3 Teacher Diet and lifestyle recommendations are nothing short of miraculous. One of the reasons I named this archetype the Teacher is that it carries an impressive wisdom of the body. Once the Teacher embarks on a program of dietary and lifestyle changes, the results are immediate. Their recovery powers are quite remarkable. I have watched cancer patients achieve extended remissions and even recovery. I have witnessed children with chronic ear infections completely end the cycle. It's all about balance.

## Problem Areas for the Teacher

However, the Teacher's gifts for analytical quickness and easy adaptation to the environment can be double-edged swords. When they are in poor health, their mental balance is upended. Usually at that point the Teacher is looking for something to turn his or her nervous system "on" and something else to turn it "off." Teachers can suffer from compulsive-type diseases if their stress hormones are not balanced with calming exercises and the right diet. This can often cause Teachers to obsess over insignificant issues and trivialities. In those circumstances, their problems can be heightened by resorting to drugs, nicotine, caffeine, and alcohol.

Tolerance can be an admirable quality, but not when it is expressed at the expense of one's own well-being. Teachers who have exhausted their immune systems with overwork, sleep deprivation, stress, or poor diet will have trouble fending off chronic infections. They become easy prey for the latest bacteria making the rounds in their offices or schools.

Teacher children suffer from chronic ear and respiratory infections, and they often appear sickly.

In general, the Teacher's immune system is rather slow to activate, particularly against bacteria and parasites. This has its benefits, since the immune system does not spin its wheels hunting down and destroying things that it could just as well have gotten along with, and this trait was critical to allowing the early Teachers to migrate throughout the world. However, the tolerant Teacher immune system can easily overidentify with the external world, lowering its defenses against microbes, harmful foods, and aberrant cells. Poor immune surveillance means vulnerability to infections and a higher-than-average risk for many common cancers.

## The Teacher's Metabolic Profile

Teachers are generally ectomorphs or meso-ectomorphs, possessing a low body-fat percentage, small bone size, a high metabolism, and a wiry physique. A characteristic of the Teacher is clearly discernible tendons and ligaments underneath the skin, a sign of flexibility and strength. Another trait is the high incidence of Teachers who have a longer index finger on one hand and a longer ring finger on the other, a sure sign of asymmetry. Teachers tend toward small to average height, with torsos and legs of about equal length. Reading this may come as a surprise if you remember that I listed Abraham Lincoln as an archetypal Teacher. But Lincoln had a secondary pituitary disease that caused his great height. However, he did have the great sinuous strength of the Teacher. It was said that he would routinely amaze his friends by holding out an ax with one hand parallel to the ground for minutes at a time.

Most Teachers are PROP "tasters" or "super-tasters," and most have an "A antigen" blood type (blood types A1, A2, A1b, A2B). Their Rh blood group is interesting. They are almost always Rh-positive, and a little further detective work shows that they almost always have the "ancient" form of the Rh-positive blood group (CDE).

Many Teachers have a significant number of whorl-type fingerprints, a useful indicator of future cancer risks, especially for women. Research has shown that the presence of six or more whorl fingerprint patterns is associated with an increased risk of breast cancer, with a statistical significance no different from that of a positive mammogram. That's pretty dramatic! In type A Teachers, a whorl count greater than six should be considered a wake-up call for a proactive cancer-prevention program. Don't worry: The Teacher GenoType Diet and lifestyle program will do that for you automatically.

## The Teacher Immune System Profile

The tolerant Teacher immune system can easily overidentify with the external world, lowering its defenses against microbes, harmful foods, and aberrant cells. Poor immune surveillance means vulnerability to infections and a higher-than-average risk for many common cancers. This often occurs in Teachers because tumor cells often "turn off" genes that are supposed to keep them in check. These "jailkeeper" genes, called tumor-suppressor genes, are supposed to prevent cancer genes from activating. Often in Teachers, the first thing cancer genes do is figure out a way to turn off the suppressor genes. At that point, the inmates are running the prison. Fortunately, with the Teacher GenoType Diet, you can put those suppressor genes back to work for you.

The wild bird that best personifies the character of the Teacher is the crane. In addition to being great fishing birds, cranes perch for lengths of time on one leg. There is a rather well-known stance in martial arts known as "crane on a rock." The practitioner balances on one leg and, like the crane, remains still, perhaps awaiting the movement of its prey before responding. As with the crane, the Teacher is a success story in acceptance and integration, highly beneficial traits for both man and environment.

## Diet and Digestion

Their ancient heritage as farmers has genetically disposed Teachers to metabolize a wide variety of fresh foods, grains, and seafood. Conversely, they lack the enzymes to digest and metabolize animal fat properly. The ideal Teacher Diet is vegetable- and seafood-based, with small amounts of other low-fat proteins. Unlike the other GenoTypes, which gain weight from excess calories or inadequate exercise, Teachers gain weight from excess toxicity. When their diets are excessively meat-based, they gradually develop a buildup of bacteria in the digestive tract, which can act as a powerful block on their metabolism and immune system. The result is a range of stomach and intestinal problems, including gastritis, which causes extreme discomfort in the upper abdomen, nausea, and in severe cases dark, bloody stools.

## Teacher Diet Dos and Don'ts

The Teacher GenoType Diet is a plant-based, low-bacteria-overgrowth, high-phytonutrient diet. *Teacher Dos* are superfoods and supplements that epigenetically reprogram their tolerant genes, help to keep their anti-cancer defense in top shape, and help optimize the metabolism, leading to increased energy and optimal weight.

### Teacher Diet Dos

The best superfoods for the Teacher GenoType have nutrients in them that:

- **Have the nutritional building blocks needed for genetic improvement.** These superfoods are rich in the gene-methylating nutrients such as vitamin $B_{12}$, choline, and the amino acid methionine.
- **Increase muscle mass and decrease body fat.** This improves the metabolic rate, which speeds up proper weight loss. The prime foods for Teacher weight loss are identified by a diamond ($\lozenge$) icon.
- **Minimize bacteria overgrowth:** These foods do not leave a carbohydrate residue in the digestive tract; thus, they do not "feed" bad bacteria.
- **Keep the immune system vigilant and in top shape.** Teachers need to keep their anticancer genes active. Usually this involves keeping their

"tumor-suppressor genes" in tiptop shape. When working properly, these genes can help keep other cancer-inducing genes silent and inactive.

- **Clean up carcinogens and mutagens from the body.** Cleaning up carcinogens requires a well-functioning liver and effective scavenging of carcinogens by special cells in the body called *macrophages* (Latin: "big eater").

## Teacher Diet Don'ts

*Teacher Don'ts* are foods that are best minimized or outright avoided. The *Teacher Don'ts* exclude from the diet those foods that:

- **Encourage microbial growth.** Many foods containing simple sugars and starches encourage bacterial overgrowth in the Teacher's gut. This puts a strain on the Teacher's overall immune efficiency.
- **Have an undesirable ratio of good to bad fats.** An undesirable ratio of omega-6 to omega-3 fats can slow down your metabolism and interfere with the proper functioning of your immune system.
- **Block your anticancer defenses.** Teachers rarely require hormone therapy. More often, they require that something be done to restore sensitivity to their own hormones. Foods that interfere with optimal hormone function in the Teacher need to be avoided, especially in the early stages of the diet.
- **Inhibit your metabolism.** Surprisingly, many of the foods that encourage bacterial overgrowth or block your genetic anticancer defenses also inhibit your metabolism and cause weight gain.

Some foods on the *Teacher Don'ts* list need only be avoided for a short period of time so that you can regain your balance. After 3–6 months, you can reintroduce them back into your diet in modest amounts. These foods are identified by a black dot (•) icon. Of course, if you are battling an illness or feel your weight beginning to creep back up, you may want to ramp up your compliance by avoiding these foods again for a while.

## Unlisted Foods

Unlisted foods are foods that don't appear to do much good or bad. They are essentially neutral, and can be used judiciously (2–5 times weekly). The nutrients in them will benefit you, but they won't specifically help you restore balance to your genes or health to your cells. Feel free to eat them—but don't neglect the foods I recommend. The GenoType Diet is always evolving and I'm frequently adding new foods, so check in with my Web site (www.genotypediet.com), especially if you have questions about a specific food.

Well, Teacher, now is the time to put word and intent into action. From here you'll skip to Chapter 14 to learn how you can get the most from the GenoType Diets. After that, we're off to Chapter 17 and the Teacher food, supplement, and exercise prescriptions.

# CHAPTER ELEVEN

# GenoType 4: The Explorer

**M**uscular and adventurous, the Explorer is a biological problem-solver with an impressive ability to adapt to environmental changes and a better-than-average capacity for gene repair. The Explorer's vulnerability to hormonal imbalances and brain strain can be overcome with a balanced diet and lifestyle.

| Typical Features of the Explorer | | |
| --- | --- | --- |
| Psychological | Biometric | Biochemical |
| • "Lateral thinker"—concerned with changing concepts and perception<br>• "Visual simultaneous"—scans several sensory inputs simultaneously<br>• Quirky<br>• Great entrepreneur<br>• Above-average intelligence | • Asymmetrical fingerprint patterns from one hand to the other<br>• Index fingers or thumbs of each hand often have different fingerprint patterns.<br>• Often left-handed<br>• Index-to-ring-finger-length ratios are often reverse of what is expected for gender (i.e., longer ring fingers in women, longer index fingers in men).<br>• Often muscular (mesomorphic)<br>• Broad-headed, often with "chiseled" facial features<br>• Square-shaped jaw and face (narrow gonial angle)<br>• Often have spade-like incisors<br>• Torso usually longer than legs | • "All-purpose"—can be virtually any ABO blood type<br>• Often non-secretor<br>• Often Rh-negative<br>• Often PROP/PTC "super-tasters"<br>• Often borderline anemic<br>• Sensitive to caffeine, fragrances, and medications |
| Superstar Explorers | • Julius Caesar (Roman dictator)<br>• Charlie Chaplin (actor, comedian)<br>• Joan of Arc (French saint)<br>• Malcolm X (political activist) | |
| Slogan | "I'll do it my way." | |

| Typical Features of the Explorer *(continued)* | |
|---|---|
| **Strengths to Count On** | • Good gene repair and illness-recovery capacity<br>• Physical stamina<br>• Great memory retention in advanced age<br>• Effective problem-solver<br>• Ages gracefully |
| **Weaknesses to Watch Out For** | • "Canary in the coal mine": environmental and chemical hypersensitivities<br>• Accident-prone<br>• Liver detoxification can be ineffective<br>• Medically hard to diagnose<br>• Tendency for blood irregularities |
| **Health Risks** | • Type 1 diabetes<br>• Anemia<br>• Autism, dyslexia, epilepsy<br>• Breast cancer (especially if female, type A blood, and left-handed)<br>• Food and environmental allergies<br>• Liver problems<br>• Yeast infections |

When I first met Simone, she was wearing a high-tech N95 face mask and I couldn't help thinking she looked like a bank robber. In fact, in a moment of misplaced humor one of the clinic residents thrust up his hands and said, "Don't shoot" as if he were anticipating a holdup. "Very funny," she dryly responded. "However, you would not feel so jolly if you had to wear this thing day in and day out just to get around."

It soon became apparent to me that Simone, a sixty-three-year-old female African American, had already seen just about every allergist, immunologist, homeopath, and nutritionist on the eastern seaboard of the United States for her crushing chemical and environmental sensitivities. She showed up at my clinic as a last resort, complete with two very large manila file folders of medical records containing every allergy panel I had ever seen—and even a few that I'd never heard of.

A bright, articulate woman, Simone and her husband were classic rags-to-riches entrepreneurs. Starting as pants pressers, they had built a chain of seven dry-cleaning establishments over the last three decades. However, about ten years earlier, Simone had noticed that she could no longer go to work without developing pounding headaches and dizziness. Fortunately, her husband and six daughters were able to keep the business going, but Simone's sensitivities soon began to take on more dangerous manifestations. A henna hair treatment five years later landed her in the hospital with life-threatening anaphylactic allergic shock. Her eyes would swell, her throat would close up, and she would have great difficulty breathing if she came in contact with anyone wearing fragrances or rooms that had been freshly painted or recently carpeted.

Her family history was unique, to say the least. Her older brother had almost been killed by a course of sulfa-type antibiotics and had been told "to avoid certain beans." Her father had died from a very severe blood disease called aplastic anemia. Her mother was still going strong at ninety-one, living by herself, playing the saxophone, and refusing to let anyone else clean her house.

I used to dread seeing patients like Simone. They'd been to countless doctors, tried virtually every imaginable approach, and now it was my turn. Years back, I would joke to my students that these were the type of patients that made you insist on a back door to your clinic—so you could run away.

But I don't feel like running away from patients like Simone anymore. After doing her blood typing and physical measurements, it was clear as day that Simone was a GT4 Explorer. I asked Simone if she had any sensitivity to coffee. "I wouldn't know," she replied, "I never drink the stuff. I used to drink tea, but I can't anymore. Keeps me up all night." And perfume? Forget about it!

After I explained a little bit about the Explorer GenoType, Simone and I went to work. One major problem we had was trying to figure out how to "deprogram" her from all the terrible nutritional advice she had received over the years. So many people had tried to help her by putting her on more and more restrictive diets that she was now terrified to eat almost anything. I explained that it wasn't the foods that were causing the problem, but rather her out-of-control reactions to these foods and other environmental chemicals. Living the rest of her life in a plastic bubble was not going to give her the nourishment to heal—we needed to bring her detoxification pathways back into balance.

Slowly, with proper diet and supplements, Simone was able to function at ever-increasing levels. Her energy, never robust to begin with, began to rise. Blood work showed that her red cell counts, although not terribly low beforehand, were now smack in the middle of normal. At her second visit, Simone exclaimed, "You can knock me over with a feather. All those other doctors telling me what not to eat, and here the only thing that worked was a diet that told me what to eat. And to think I was almost too scared to try it."

Within two months, the mask came off and she presented it to me as a parting gift on her third and last visit. I keep it in my workshop to this day to use when I'm painting!

## Explorers at Their Best

Explorers often enjoy greater longevity than the other GenoTypes. Many of the genes we typically find in Explorers, such as the Rh-negative blood type, are common in areas of the world where people seem to live forever—such as the Basque provinces of Spain and the Caucasus Mountains of Asia. Explorers can benefit greatly from the GenoType Diet and should expect to lead long and healthy lives if they follow the recommendations in this book.

I named this GenoType "Explorer" because of the unique and often unconventional ability of those of this type to search and discover who they are in the world. Although it may sound like a cliché, the phrase "think outside the box" really does apply to these folks. Perhaps it's because Explorers are modern-day descendants of "glacial refugees" who survived by finding their way through the rapidly moving ice floes of the Last Glacial Maximum 12,000 years ago. Or perhaps it is just the way that they are wired out of the womb.

## Problem Areas for the Explorer

Explorers are very often medical enigmas. They can be challenging to diagnose, since nothing apparent or obvious presents itself as a problem. Physically, they may appear to be in good health, but they will complain of a sudden loss of energy or a sudden inability to tolerate a certain food, supplement, or drug. Explorer women often suffer from chronic yeast infections or heavy periods. Blood tests often reveal anemia or other blood disorders.

Explorers often have problems with the liver or gallbladder. This can sometimes manifest as intolerance to fats or sudden breakouts on the skin. Migraines are not uncommon in Explorers.

Caffeine sensitivity is a hallmark of Explorers, because they are almost always what geneticists call "slow acetylators"—a fancy way of saying drugs spend a long time in their livers, going round and round, when they should just get processed and eliminated. Like a man who shakes his fist at the bicycle in the road that just missed hitting him—and totally ignores the bus heading his way—the liver of the Explorer GenoType will often overreact to small levels of toxins, to the point that it lets larger amounts of toxins pass by without doing anything about them.

Explorers can be accident-prone. This is probably the result of the same quirky wiring that makes them so successful at many creative and entrepreneurial pursuits. However, if you are ever a passenger in a car with an Explorer at the wheel, and you need to turn right immediately, yell "Turn left" instead. An Explorer will almost always turn right!

"Wrong Way Corrigan" was probably a GT4 Explorer. In 1938, Douglas Corrigan sought permission to take off from New York and fly to Ireland. However, permission was denied because his $299 airplane was not considered airworthy. Corrigan then took off for his home in California instead. The next day he landed in Dublin, claiming that a faulty compass caused him to fly in the wrong direction. His "error" caught the attention of the media and the legend of "Wrong Way Corrigan" was born.

## The Explorer's Metabolic Profile

The unique metabolic profile of the Explorer is manifested in very distinct physical characteristics. They are typically mesomorphs, possessing a low to medium body-fat percentage, a high metabolism, and a large amount of muscle mass and muscle size. They can be rather large-boned, and the men tend to have asymmetrical, chiseled, craggy faces. Their trunk length is usually longer than their total leg length, and their upper legs are usually longer than their lower.

Explorers tend toward asymmetry and often have different fingerprint patterns on their left and right index fingers, one often being that rather uncommon radial loop pattern. Another asymmetry often found in Explorers is that their finger lengths tend to be backward for their gender—men often having a longer index finger on one or both hands, and women vice versa.

A lot of left-handers are Explorers, as are people with Rh-negative blood type, and although almost any ABO blood type can be an Explorer, "non-secretors" are more common.

## The Explorer Immune System Profile

Explorers often have sluggish bone marrow function and struggle to keep up their white blood cell counts. This GenoType is prone to many types of anemia, such as those that result from inadequate levels of folic acid, $B_{12}$, and iron, and other types of anemia that result from bone-marrow suppression or low levels of an enzyme called G6PD. G6PD is critical to the body because it enables the production of a critical antioxidant called glutathione. In addition to powerful detoxification effects in the liver, glutathione protects red blood cells against damage caused by certain drugs and foods.

Explorers often have trouble clearing foreign or man-made chemicals from their blood. This clearing process is called *acetylation*. Efficient acetylation helps drugs become more effective and detoxifies cancer-causing substances. GT4 Explorers have problems detoxifying drugs, carcinogens, and various compounds that have been introduced into the environment by artificial means, such as pesticides, fertilizers, and hydrocarbons.

Because of these issues, Explorers can be quite chemically sensitive, and they often react negatively to "typical doses" of drugs, antibiotics, and even vitamins and minerals. When using these medicines, they should always start with the lowest doses and gradually work their way up.

## The Explorer GenoType Diet

Explorers who maintain a detoxifying diet that also nourishes the blood and bone marrow will have few health problems and usually will have very little trouble attaining a healthy weight.

If you are an Explorer, you can modify the genes that cause poor detoxification in your own lifetime—but even better, you can take steps to change the forecast for generations to come. With the right diet and lifestyle for your GenoType, you can be caretakers of both the young and the old. Perhaps like Simone, you may be surprised to find out that food sensitivities and toxicity are best treated by the proper foods for one's body, not just avoiding the wrong ones.

## Explorer Diet Dos and Don'ts

The Explorer diet could be described as "Entry-Level Neolithic," since the circumstances of its creation appear to be on the tipping point between the twilight of the purely hunter-gatherer technologies and the approaching dawn of the agricultural revolution. *Explorer Dos* are superfoods and supplements that epigenetically reprogram their reactive genes, help to keep their detoxification mechanisms in top shape, and help optimize the metabolism, leading to increased energy and optimal weight.

## Explorer Diet Dos

The best superfoods for the Explorer GenoType have nutrients in them that:

- **Have the nutritional building blocks needed for genetic improvement.** These superfoods are rich in gene-methylating nutrients such as vitamin $B_{12}$, choline, the amino acid methionine, and the histone regulators curcumin, copper, and biotin.
- **Increase muscle mass and decrease body fat.** Detoxifying Explorers improves their metabolic rate, which speeds up proper weight loss. The prime foods for Explorer weight loss are identified by a diamond ($\Diamond$) icon.
- **Build up the blood.** Explorers need nutrient-rich blood so they can maximize their oxygen-carrying capacity and keep their resistance up and energy levels in top shape. Foods rich in selenium, iron, and the B vitamin thiamine are Explorer superfoods.
- **Clean up toxins from the body.** Cleaning up toxins requires a well-functioning liver and bile ducts. Explorers can get the job done better by eating plant foods that contain polysaccharide nutrients that help support these cells. The Explorer also benefits from foods that enhance the liver's Phase I and Phase II detoxification pathways.

## Explorer Diet Don'ts

*Explorer Don'ts* are foods that are best minimized or outright avoided. The *Explorer Don'ts* exclude from the diet those foods that:

- **Contain toxins or molds.** Many grains, nuts, seeds, and dairy products can interfere with the proper detoxification function in the Explorer. The Explorer diet minimizes your exposure to pesticides and other undesirable environmental toxins.
- **Inhibit your metabolism.** Surprisingly, many of the foods that encourage toxic buildup in the Explorer also inhibit your metabolism and cause weight gain.

- **Have an undesirable ratio of good to bad fats:** An undesirable ratio of omega-6 to omega-3 fats can slow down your metabolism and interfere with the proper functioning of your immune system.
- **Contain reactive lectins or other allergens:** Lectins are proteins that can interfere with proper digestive and immune function. Phenols are plant compounds that cause allergic reactions in many Explorers.

Some foods on the *Explorer Don'ts* list need only be avoided for a short period of time so that you can regain your balance. After 3–6 months, you can reintroduce them back into your diet in modest amounts. These foods are identified by a black dot (•) icon. Of course, if you are battling an illness or feel your weight beginning to creep back up, you may want to ramp up your compliance by avoiding these foods again for a while.

## Unlisted Foods

Unlisted foods are foods that don't appear to do much good or bad. They are essentially neutral, and can be used judiciously (2–5 times weekly). The nutrients in them will benefit you, but they won't specifically help you restore balance to your genes or health to your cells. Feel free to eat them—but don't neglect the foods I recommend. The GenoType Diet is always evolving and I'm frequently adding new foods, so check in with my Web site (www.genotypediet.com), especially if you have questions about a specific food.

Well, Explorer, now is the time to put word and intent into action. From here you'll skip to Chapter 14 to learn how you can get the most from the GenoType Diets. After that, we're off to Chapter 18 and the Explorer food, supplement, and exercise prescriptions.

# CHAPTER TWELVE

# GenoType 5:
# The Warrior

**Long, lean, and healthy in youth,** the Warrior is subject to a bodily rebellion in midlife. With the optimal diet and lifestyle, the Warrior can overcome the quick-aging metabolic genes and experience a second "silver age" of health.

| Typical Features of the Warrior | | |
| --- | --- | --- |
| Psychological | Biometric | Biochemical |
| • "Choleric temperament"—charismatic, if occasionally bad-tempered<br>• Quick and nimble computer-like mind<br>• Needs to learn how to relax<br>• Will pursue a mental challenge inexorably until it is mastered | • Legs usually longer than torso<br>• Often barrel-chested in later life<br>• "Long headed"—dolichocephalic<br>• Soft, oval jawline, which tends to recede with age<br>• Slender in youth; pear-shaped or barrel-chested later in life<br>• Tends to flush when nervous or under stress<br>• Often has one or two arch-type fingerprint patterns<br>• Index-to-ring-finger ratios are usually symmetrical | • Often blood type A or AB<br>• Non-secretors and secretors<br>• Typically Rh-positive<br>• PROP/PTC non-taster or super-taster<br>• Tendency for skin to flush<br>• Tendency for blood to clot too easily |
| Superstar Warriors | • Dwight Eisenhower (U.S. president and military leader)<br>• Michelangelo's "David"<br>• Julia Child (chef, actress)<br>• Hillary Clinton (U.S. politician) | |
| Slogan | "Time flies when you're having fun." | |
| Strengths to Count On | • Ox-like strength<br>• Recovers well from illness<br>• Beautiful physical specimens In youth | |
| Weaknesses to Watch Out For | • Thrifty metabolism; stores calories as fat<br>• Ages early and steadily<br>• Stress tends to depress immune system and cause blood to thicken | |

| Typical Features of the Warrior *(continued)* | |
|---|---|
| Health Risks | • Insulin resistance and obesity in midlife<br>• Chronic digestive problems, cramps, and bloating<br>• Hormonal imbalance and infertility, often in early adulthood<br>• Heart disease, high blood pressure, stroke |

The phone call came through late Saturday night: Could I please call the rabbi as soon as possible? Returning from my younger child's school orchestra recital, I listened to the message, knowing that given the circumstances this would be no ordinary return phone call. The rabbi had done very well with naturopathic care, as readers of my first book, *Eat Right for Your Type*, might remember.

However, there was a new challenge. The rabbi's oldest son, also a rabbi, had begun to develop what appeared to be a series of mini strokes, leaving him at times confused, disoriented, and bedridden. A young man in his forties, he was the rabbi's right-hand man and destined to be successor, so this development had thrown the congregation into an uproar.

That next week, I met Mordecai at the clinic. His reputation for having a prodigious intellect had preceded him; his sister had described him as a "walking computer," with knowledge of the Jewish scripture that was considered photographic. Physically, his appearance came as a bit of a shock. The rabbinic scholars I'd met in the past were typically small and frail, with a pale complexion—the so-called indoor tan. Mordecai was a bear of a man. Tall, with a huge, luxurious graying beard and a rosy, florid complexion, he had large brown eyes that were simultaneously intense and playful, accompanied by an infectious grin. Within minutes, the staff and I had taken to calling him "The Bear."

Actually, I had smelled The Bear's presence before I had seen him. The fellow was obviously a chain-smoker—to the point that the nicotine had permeated every nook and cranny of his body and clothing. When I asked him why he continued to smoke despite the obvious health risks, he gave a typical Bear answer: Yes, he was indeed scared that the smoking would give him cancer, but he thought that in the meantime the cigarettes were probably suffocating any potential cancer cells!

One look at The Bear's head told me he was a GT5 Warrior. From the front it was normal enough; however, from the side it looked enormous! He was quite tall, but stocky, with a full chest. Ruth, his wife of twenty-five years, told me that The Bear had never been very health-conscious, eating whatever he desired. As a young man he was quite thin, but over time he had steadily gained weight, although he had never changed his eating habits. He had a high-stress lifestyle as the elder rabbi's second in command. He was forever putting out fires in the community—from finding shelter for a dispossessed widow, to counseling young married couples about Tay-Sachs disease, to raising money to repair the synagogue's leaky roof. The Bear was tireless.

However, health problems, long delayed, were coming home to roost. His blood pressure had risen steadily in the last few years and was now at alarming numbers: 200/130, despite medication. He was spilling protein in his urine, a sure sign that the high blood pressure was wearing out the delicate filtering devices in the kidneys. Recently, he had had what the cardiologist was calling "mini strokes," where he would become quite clumsy, his speech would slur, and he would suffer from double vision. He had been warned that these symptoms were a clear sign that he was at risk of developing a true stroke in the future.

I explained to both Ruth and The Bear that as a GT5 Warrior, his arteries were more prone to a type of inflammation that literally burns the sides of the blood vessels. If we could get his flaming arteries under control, perhaps we could get these mini strokes under control. Everyone in the clinic loved The Bear and we were especially touched by the graciousness of Ruth, who always showed up at the clinic with something special. Figs from Israel and spelt matzo bread were two favorites. It was soon obvious that she was the power behind the throne, and with her help we crafted an approach that allowed her to control Bear's diet and make sure that he took his supplements. I soon discovered that it was useless to talk to The Bear about making changes for his own benefit: His mentality was back in the past, where as a young man he could eat anything with impunity. However, with the help of Ruth we soon discovered that you could get The Bear to do almost anything if you framed it in a way that made it appear that other people would be hurt if he continued to resist changing his lifestyle.

Soon things began to improve. The Bear began walking every day. He lost weight. His cholesterol began to drop. At six months into the program, he started using the nicotine patches, and within two weeks he had given up smoking completely. He was now sleeping well, something his wife had never observed before. Up until now, his sleep had been so restless that she had jokingly referred to him as "the thrasher."

Sadly, The Bear's father, the elder rabbi, has since passed away. Now the head of the congregation, The Bear has more, not less, responsibility and stress. Yet the members of the synagogue (many of whom have become patients as well) say that he is mentally sharper and more acute now after his illness than he was before.

## Warriors at Their Best

The Warrior is all about youth. Warriors emerge from the womb strong and healthy, and are beautiful children, with few health problems and a natural fitness that is the product of enviably efficient metabolisms. Young Warriors have high-growth-factor activity levels, which give them long legs with strong, visible tendons and ligaments.

As teenagers, Warriors are usually quite attractive. The men are often "beautiful" in an almost androgynous way, while the women exude an alluring quality that is hard to define. Their inquisitive minds and budding youthfulness often show up as an intense vibrancy in their facial features.

Warriors have soft, oval faces but do not carry extra padding under their skin. In fact, although they have ravenous appetites when young, they are often so slim that they are underweight.

The Warrior GenoType usually has a quick and nimble mind with a tremendous capacity for memory retention. Like the Teacher, they know how to get to the gist of an issue, the heart of the matter. In early adulthood, they are not only charismatic and attractive to the opposite sex, but they are known to be among the most fertile of the GenoTypes.

## Problem Areas for the Warrior

This happy time for the Warrior persists throughout adolescence and young adulthood, but as they approach midlife, their formerly efficient metabolic genes seem to "hit a wall" and almost stop working. These changes begin imperceptibly at first, but in a relatively short time, the aging process speeds up.

I rarely see Warriors in my clinic before middle age. They're just too busy, building their lives, raising their families, and pursuing their careers. They often forgo checkups—why bother, since they're usually in such good health? Like The Bear, in their younger years Warriors don't usually pay much attention to diet and exercise. Young Warriors are the people who can drop ten pounds with little trouble.

But as they hit their late thirties, their thrifty epigenetic inheritance starts to kick in and their aging process accelerates. By midlife, Warriors are having trouble losing weight. Their waistlines gradually thicken, accumulating the more destructive belly fat.

By their forties they're usually complaining, "I don't look like myself anymore." Warriors begin to see more and more flesh when they look in the mirror. The neck area becomes thickened, softer, and less defined. Warriors with especially wide jaw angles may find their jawlines gradually receding. BMI and waist-to-hip ratio rise steadily over time, especially if they have high-fat and high-sugar diets and don't exercise. Their previously strong, slender bodies begin to sag and fat accumulates around the waist area. In both men and women, hair thins and bags appear under the eyes.

The circulatory system is the Achilles' heel of the Warrior. Programmed for overactivity almost from the moment of their conception, the Warrior's unique cardiovascular system is the source of their health problems: from the tendency to develop hemangioma-type "port-wine stains" and "angel kisses" on their skin in early age, to a tendency in early middle age to flush when they are stressed out, to problems with their blood pressure and heart in later middle age.

The Warrior can also have problems controlling the viscosity, or thickness, of their blood. This is especially true of Warriors when they are under intense and prolonged stress.

An unhappy trend in later life for many Warriors is a winding down of their bowel function. Stools become more compact, difficult to eliminate, and infrequent. Chronic constipation, cramps, gas, and bloating are partially caused by a weakening of the abdominal muscles and the extra pressure they place on internal organs. Fortunately, these symptoms can be relieved with the right diet and exercise program.

## The Warrior's Metabolic Profile

Although the Warrior worldview is essentially thrifty, unlike the Gatherer GenoType the Warrior is not a reaction to environmental scarcity but rather a result of the founder effect as applied to many of the emerging civilizations, beginning with the Neolithic Era (11,000 years ago) and ending by the Iron Age (2,500 years ago). The almost constant warfare, plague, and subsistence nutrition during these times had major epigenetic effects on the population (especially in Europe), and the Warrior is a survivor of this.

Young Warriors have high-growth-factor activity levels, which give them long legs with strong, visible tendons and ligaments. Female Warriors are particularly long-legged. They have soft, oval faces, but unlike earlier thrifty genetic types, Warriors are not rounded and do not carry extra padding under their skin. In fact, although they have ravenous appetites, they are sometimes underweight. Their thriftiness is an epigenetic response to urgency, not scarcity.

Like the Teacher, the Warrior is a GenoType of the A-blood-group antigen (blood types A and AB). Warriors have a funny relationship with the ability to taste PROP. They either can't taste it (non-taster) or it is so unpleasant that they just want to spit out the test strip (super-taster).

Warriors tend to have long heads—what could almost be described as an egghead. The Warrior head is longer than it is wide. Studies have shown that modern humans have two distinct physical traits that differ from our ancient and medieval ancestors: We are taller and our heads are more elongated. The Warrior usually has both of these traits to some degree. Warriors are often tall and long-legged and, because of this, can carry a surprising amount of weight on their frames without looking obese.

The arch fingerprint pattern is a hallmark of the Warrior and is often seen on the thumbs and index fingers.

Warriors are usually fast-acetylators. Unlike the GT4 Explorer, who is an underacetylator and can have trouble detoxifying environmental chemicals, the GT5 Warrior often creates his or her own unique problems by overacetylating environmental toxins. For this reason, red meat is not a great food choice for Warriors—they convert the by-products found in well-cooked red meat into carcinogens that bind to DNA and cause cancers of the stomach, colon, and breast. The combination of their thrifty epigenetics and the fast-acetylator polymorphism also makes the Warrior a prime candidate for adult-onset diabetes.

## The Warrior Immune System Profile

Warriors usually have great immune surveillance. They don't get sick for long periods of time, and usually when they do they have better than average rates of recovery.

Warriors are often the patients who beat cancer or survive heart attacks. They are able to metabolize drugs and even toxins fairly easily, and they are not all that prone to allergies. This genetic heritage harkens back to ancient times. Warriors were among the first urban survivors, capable of surviving plagues and pestilence that knocked the less successful GenoTypes out of the ring.

## The Warrior GenoType Diet

The Warrior evolved as a farmer-agrarian GenoType, so it seems almost counterintuitive that they made some of the finest warriors in history. We tend to picture GenoTypes such as the Hunter as the typical warrior, but this is simply not true; most hunters were not warriors. Their mission was chasing prey, not fighting. Since they owned nothing, they had nothing to defend. Farming societies cultivated the best warriors for the simplest of reasons—they were well-organized and had tangible assets and loved ones to defend. Thus, the ideal diet for Warriors is a Mediterranean-style diet that incorporates a mix of fish, oils, grains, vegetables, and fruits.

Many of my patients fear that by the time the camera captures their sagging jawlines and spreading waistlines, it's too late to turn back. I tell them, "You're not a kid anymore, but you can become a fit and youthful forty- or fifty-year-old." It just takes determination and the right plan. If you are a Warrior GenoType, with modest effort you can get back into great shape and give yourself a second "silver age" of health and vitality. Perhaps not as invulnerable as the first, but pretty darn good—maybe even better, actually, since it comes with all that life experience.

After all, wasn't it George Bernard Shaw who said, "Ah the pity that advice, like youth, is wasted on the young"?

## Warrior Diet Dos and Don'ts

The Warrior GenoType Diet is a "modified Mediterranean," plant-based, low-glycemic, high-phytonutrient diet. *Warrior Dos* are superfoods and supplements that epigenetically reprogram their thrifty genes, helping to slow down the aging process, and help optimize the metabolism, leading to increased energy and optimal weight.

## Warrior Diet Dos

The best superfoods for the Warrior GenoType have nutrients in them that:

- **Have the nutritional building blocks needed for genetic improvement.** These superfoods are rich in the gene-methylating nutrients such as vitamin $B_{12}$, choline, and betaine, and the histone regulators curcumin, copper, and biotin.
- **Rich in hormone-balancing lignans.** Lignans are sometimes referred to as phytoestrogens. However, this term is misleading. Many of the disease-fighting properties of lignans have nothing to do with any type of hormone activity. Lignans appear to offer the Warrior protection against many diseases, including hormone-dependent breast and prostate cancer, osteoporosis, cardiovascular disease, and inflammation. Major source of lignans are flaxseed, sesame seeds, and many fruits.
- **Increase muscle mass and decrease body fat.** Balancing the hormones of the Warrior improves their metabolic rate, which speeds up proper weight loss. The prime foods for Warrior weight loss are identified by a diamond (◊) icon.
- **Protect and nourish the arteries and heart.** With age, many Warriors develop problems with their circulation. Foods rich in antioxidants called *flavones* help protect the delicate lining of the arteries from damage. Many of these flavones are found in tea, fruits, and vegetables.
- **Slow down the biologic clock.** Other antioxidants, such as resveratrol, vitamin E, and selenium, may help control accelerated aging in the Warrior GenoType.

## Warrior Diet Don'ts

*Warrior Don'ts* are foods that are best minimized or outright avoided. The *Warrior Don'ts* exclude from the diet those foods that:

- **Burn the arteries.** Red meats and all sources of trans fats should be avoided. They cause arterial inflammation, thicken the blood, and accelerate the aging process in Warriors.

- **Inhibit your metabolism.** Surprisingly, many of the foods that cause hormonal imbalance in the Warrior also inhibit your metabolism and cause weight gain.
- **Have an undesirable ratio of good to bad fats.** An undesirable ratio of omega-6 to omega-3 fats can slow down your metabolism and interfere with the proper functioning of your immune system.
- **Are high-glycemic foods.** High-glycemic foods produce large fluctuations in blood glucose and insulin levels. Avoiding these is the secret to reducing your risk of heart disease and diabetes and is the key to sustainable weight loss.

Some foods on the *Warrior Don'ts* list need only be avoided for a short period of time so that you can regain your balance. After 3–6 months, you can reintroduce them back into your diet in modest amounts. These foods are identified by a black dot (•) icon. Of course, if you are battling an illness or feel your weight beginning to creep back up, you may want to ramp up your compliance by avoiding these foods again for a while.

## Unlisted Foods

Unlisted foods are foods that don't appear to do much good or bad. They are essentially neutral, and can be used judiciously (2–5 times weekly). The nutrients in them will benefit you, but they won't specifically help you restore balance to your genes or health to your cells. Feel free to eat them—but don't neglect the foods I recommend. The GenoType Diet is always evolving and I'm frequently adding new foods, so check in with my Web site (www.genotypediet.com), especially if you have questions about a specific food.

Well, Warrior, now is the time to put word and intent into action. From here you'll skip to Chapter 14 to learn how you can get the most from the GenoType Diets. After that, we're off to Chapter 19 and the Warrior food, supplement, and exercise prescriptions.

# CHAPTER THIRTEEN

# GenoType 6: The Nomad

A GenoType of extremes, with a great sensitivity to environmental conditions—especially changes in altitude and barometric pressure—the Nomad is vulnerable to neuromuscular and immune problems. Yet a well-conditioned Nomad has the enviable gift of controlling calorie intake and aging gracefully.

| Typical Features of the Nomad | | |
| --- | --- | --- |
| Psychological | Biometric | Biochemical |
| <ul><li>"Phlegmatic"—easygoing, roll-with-the-punches personality</li><li>Quiet but witty</li><li>Generally optimistic, rational, and fun-loving</li><li>Tends to keep emotions hidden</li><li>Ability to utilize visualization for health and recovery</li></ul> | <ul><li>White lines throughout the fingerprint</li><li>High frequency of ulnar loop fingerprint patterns</li><li>Physically symmetrical</li><li>Index and ring-finger lengths are usually symmetrical to gender</li><li>Extremes of height—short or tall</li><li>Squarish head shape</li><li>Legs usually longer than torso</li><li>Waist-to-hip ratios high in men, low in women</li><li>Small teeth; incisor shoveling common</li><li>Higher-than-average number of redheads, green eyes</li></ul> | <ul><li>Blood type B and AB</li><li>PROP taster</li><li>Almost always Rh-positive</li></ul> |
| Superstar Nomads | <ul><li>Elizabeth I (English monarch)</li><li>Peppermint Patty (Peanuts cartoon character)</li><li>Abraham (biblical patriarch)</li><li>Winston Churchill (British Prime Minister)</li></ul> | |
| Slogan | "A new career in a new town." | |
| Strengths to Count On | <ul><li>Great mind-body connection</li><li>Balanced immune system—when healthy, not prone to allergies or inflammation</li><li>Good stress-handling abilities</li></ul> | |

| Typical Features of the Nomad *(continued)* | |
|---|---|
| **Weaknesses to Watch Out For** | • Extremes and variables make diagnosis difficult<br>• Sensitive digestive tract—can be gluten-intolerant<br>• "Garbage removal" by the immune system can become compromised |
| **Health Risks** | • Tendency toward "slow infections" such as long and lingering viral disorders, warts, or parasites<br>• Neuromuscular diseases with age<br>• Fatigues easy<br>• Alzheimer's disease |

I sometimes have difficulty relating to adolescent patients because I sense they've been dragged to the clinic by their parents. Many of them simply go passive-aggressive on me, staring at their shoes, providing one-syllable answers, sighing in exasperated tones as only teenagers can do. We doctors don't often make things much better—usually relating to the child through the intermediary of the parents, or trying to motivate these kids with abstract concepts like "health." Most doctors have forgotten what it is like to be a kid. Unlike adults, kids relate to the world in very concrete terms—just observe how a school guidance counselor talks to a teenager and you'll see my point.

Yet Claire was different from your average teenage patient. For openers, she had obviously seen a lot of doctors for her history of crushing fatigue. They were mostly pediatricians, but also included an immunologist and an infectious-disease specialist. She had also seen a few nutritionists. Yet despite all this work and worry, Claire was a very engaging, charming eighteen-year-old with loving parents and siblings.

However, other than a tentative diagnosis of chronic fatigue syndrome and Lyme disease, nobody had been able to figure out what was wrong with the young woman. Technically a sophomore in high school, Claire had not been to class since her health problems started almost three years earlier. Instead, she was schooled by tutors who visited the house a few times a week. However, even a light, abbreviated workload exhausted her and she was very worried about falling further and further behind in her studies.

Physical examination disclosed that Claire had swollen glands under her chin and along the neckline. She also had a slightly enlarged spleen. As a baby, Claire had a condition called "premature thelarche," the early appearance of breast development in females, before the normal age. Her blood type was B and she was a non-secretor. She had very low blood pressure and would often get quite dizzy if she stood up rapidly from a sitting position. Claire had ten ulnar loop fingerprint patterns, and every fingerprint was shot through with "white lines"—areas where the pattern did not print because the fingerprint ridges had been worn away. She had two very long index fingers.

Claire's mom confided in me that "Claire was like two different people, depending on the weather." A drop in barometric pressure, typically a sign of an impending storm, would send Claire to bed with blinding headaches. When it was a clear and sunny day, she had a bounce in her step.

GenoTyping Claire showed that she was clearly a GT6 Nomad, from the high frequency of ulnar loops, to her B blood type, to the great sensitivity to weather and barometric pressure. Quite a few Nomads suffer from chronic fatigue syndrome, and the cause and cure of this, as well as most of Claire's myriad health problems, boils down to a tiny molecule called nitric oxide.

Made from only one atom each of nitrogen and oxygen, and so ephemeral that it disappears almost as fast as it is made, nitric oxide (usually abbreviated NO) had escaped the attention of medical researchers up until only a few years ago—simply because nobody knew it was there or how to find it. Yet NO does a slew of incredibly important things in the body. It helps activate cells called macrophages, cells of the immune system that get rid of "bad" debris during injury and illness. As the "garbagemen" of the immune system, macrophages use nitric oxide to sweep up parasites, viruses, and other infectious rubbish so it can be eliminated.

Like Claire, many GT6 Nomads struggle with keeping their NO at healthy levels and evenly distributed in their bodies. Often they can have excessive amounts in some areas and be lacking in others. In Claire's case, there was no shortage of NO in her circulatory system—that was what was causing the low blood pressure and perhaps the sensitivity to weather. However, other parts of the body—the immune system and nervous system, for example—clearly appeared starved of this vital nutrient.

Fortunately, NO regulation can be enhanced by diet and supplementation. We started Claire on the GT6 Nomad Diet and added several supplements to her protocol, in addition to advising her parents to discontinue the more than forty individual supplements she was taking daily on the advice of her other physicians and nutritionists. I also had Claire begin to do visualization exercises—to imagine herself full of vibrant energy, with all the systems of her body working together in complete harmony.

It took some time—about six months—but gradually Claire began to turn around. She loved doing the visualization exercises, telling me that it made her feel "centered and relaxed." Within a few weeks, she started to grocery-shop with her mother and soon was out in the evenings walking with the family dog. As a young child, she had enjoyed working with her father on his 1968 Mustang, and just about the time she returned to school that fall, she had been able to resume their special hobby.

I'd thought I'd never see Claire again when her family relocated to Texas because of her father's job promotion. But just the other day, I received a sweet note and a cute prom picture of Claire and her handsome, if somewhat excessively tattooed and body-pierced, boyfriend.

"Wouldn't have been here without you. Love Claire" was all it said.

## Nomads at Their Best

Everyone should have a Nomad as a friend, including other Nomads. This GenoType has an easygoing, roll-with-the-punches personality. Nomads are generally optimistic, rational, and fun-loving.

You might say that the Nomad is the "great communicator." This is not just a psychological characteristic but one that infuses every aspect of their physiology. Their uncanny ability to control nitric oxide activity is at work with their ability to think or visualize their way to better health. Nomads are the patients who get dramatic healing results from meditation and mental exercises—the ultimate example of a mind-body connection.

Overall, Nomads have excellent metabolic capacity and in a state of proper health and balance are not prone to obesity, diabetes, or cardio-vascular disease. They usually have normal hormone function, tend to have little problems with stress, and sleep restfully.

## Problem Areas for the Nomad

Loss of proper nitric oxide control can interfere with healthy aging for Nomads. This ability must be sustained by diet and lifestyle: A break-down of this important Nomad function is almost always accompanied by problems with the immune system, circulatory flow, and brain func-tion as they age.

Nomads have a tendency to develop slow-growing neurodegenerative diseases, triggered by viral infections in youth that don't show up until much later in life. They seem to have a higher than average rate of autoim-mune diseases like lupus, multiple sclerosis, and ALS ("Lou Gehrig's disease").

The liver and spleen can be problem areas for the Nomad, and there-fore this GenoType has more than its fair share of inflammatory liver dis-ease, hepatitis, and cirrhosis.

## The Nomad's Metabolic Profile

Nomads tend to be large-boned, and their BMI and waist-to-hip ratio are often higher than average; this isn't necessarily an indication of excess fat, but rather muscularity. Nomads are often mesomorphs, possessing a low to medium body-fat percentage, medium to large bone size, a medium to high metabolism, and a large amount of muscle mass and muscle size. Shorter Nomads have lankier builds, with less muscular necks and upper bodies. Most tall Nomads have broad, balanced facial and nose shapes, and many have squarish jaws. Shorter Nomads tend to have more even features and rounder jawlines.

The fetal development of the Nomad seems to be heavily influenced by altitude. In high-altitude environments Nomads will often be tall, while at lower altitudes they will be shorter. Shorter Nomads are usually more asymmetrical, while taller Nomads tend toward a more symmetrical appearance. A common sign of Nomad asymmetry is a difference in the size of the breasts in women and the testicles in men, whereas a common sign of symmetry is that their index-to-ring-finger ratios usually behave as expected—that is, male Nomads having longer ring fingers on both hands and female Nomads longer index fingers. Almost all taller Nomads tend to be symmetrical—a sure sign that they enjoyed life in the womb.

The Nomads' fingerprint dermatoglyphics are often distinctive. They often have an abundance of ulnar loop patterns, and if the total ulnar loop count is higher than eight to ten and there is a history of Alzheimer's disease in their family line, they might very well benefit from following the suggestions found in the Supplement Section of the Nomad Diet Plan to maintain cognitive and memory skills in later life.

## The Nomad Immune System Profile

Like just about every other aspect of Nomads, their immune system is a study in extremes. They were early herders, and this allowed for their rapid and almost continuous migration and exposed them to widely varying climates, flora, and fauna. The first societies to adopt a horseback existence were Nomad GenoTypes, a technological breakthrough that transformed the human relationship with time and space. Migration could now extend over hundreds of miles, rather than simply up the mountain one season and down the valley the next. This may be the reason for the Nomad's overly tolerant immune system. Unlike the Teacher, who essentially walked from one place to another and thus had more time to adjust to their new homes, the mounted Nomad could quickly come into contact with a variety of new dangerous microbes. This is exactly the type of problem that concerns today's infectious-disease specialists: airline passengers who can hop on a plane, pick up a tropical disease, and return home to infect neighbors with no natural resistance.

This gave the Nomad a more specific, idiosyncratic tolerance than the Teacher, and an immune system likely to struggle with low-grade, chronic viral infections, many of which can linger for life. If their nitric oxide production is unbalanced, their immune system will be sluggish when it comes to attacking and clearing these invaders. A sure sign of this in the Nomad is a crushing, numbing fatigue.

Yet in some Nomads the immune scenario will be, if anything, overactive. If this is the case, the cause is almost always an excess of activity of the so-called killer cells of the immune system. When this is the case, the Nomad can suffer from autoimmune diseases such as lupus, rheumatoid arthritis, and sarcoidosis. This scenario is especially common in Nomads of African or Asian ancestry and Nomads who have white lines throughout their fingerprint patterns.

## The Nomad GenoType Diet

Nomads present a mixed dietary picture, and it requires a little extra work to get the balance right. They tend to display certain sensitivities, especially to proteins called lectins, which are present in many foods. Some Nomads are also sensitive to gluten, as evidenced by white lines on their fingerprints. These variations make the Nomad Diet more idiosyncratic than most.

Nomads are one of the rare GenoTypes genetically adapted to fermented dairy products in the diet, although there are some Nomads who are lactose-intolerant. This adaptation by herding and milking societies allowed people to continue consumption of abundant dairy foods throughout their lives. A sure sign that Nomads have had herding in their genetic history is the presence of incisor shoveling, a grooving of the back surface of the upper front teeth. Another sign of adaptation to milking societies is the Nomad's tendency to have smaller teeth.

As a Nomad GenoType, you possess many natural gifts. And while being "special" can sometimes feel frustrating because you don't fit neatly into prescribed patterns, if you find your own way, you can be hale, healthy, and wise late into life. The GenoType Diet is all about the science of individuality, and Nomads are true individuals.

## Nomad Diet Dos and Don'ts

The Nomad GenoType Diet is a "herder diet"—an omnivorous, low-lectin, low-gluten diet. *Nomad Dos* are superfoods and supplements that epigenetically reprogram their thrifty genes, helping to slow down the aging process, and help optimize the metabolism, leading to increased energy and optimal weight.

### Nomad Diet Dos

The best superfoods for the Nomad GenoType have nutrients in them that:

- **Have the nutritional building blocks needed for genetic improvement.** These superfoods are rich in gene-methylating nutrients such as vitamin $B_{12}$, green tea, and histone regulators such as curcumin, ginseng, sage, and biotin.
- **Optimize the production and regulation of nitric oxide.** Keeping the Nomad cardiovascular, immune, and nervous systems amply supplied with nitric oxide enhances their total function, since in Nomads the function of the three as a whole is greater than the sum of the parts. For the Nomad, foods high in the amino acids arginine and citrulline are NO superfoods.
- **Rebuild the gut lining.** Nomads often have very compromised gut linings, often evidenced as white lines on their fingerprints. Foods rich in short-chain fatty acids and probiotics (friendly bacteria) will soon help the digestive tract rebuild itself.
- **Increase muscle mass and decrease body fat.** Optimizing nitric oxide production in Nomads improves their metabolic rate, which speeds up proper weight loss. The prime foods for Nomad weight loss are identified by a diamond (◊) icon.

## Nomad Diet Don'ts

*Nomad Don'ts* are foods that are best minimized or outright avoided. The *Nomad Don'ts* exclude from the diet those foods that:

- **Cause hypoglycemia.** Since low blood sugar can cause such debilitating fatigue in Nomads, you will want to maintain a diet that keeps your blood sugar levels within a narrow, optimal range.
- **Inhibit your metabolism.** Surprisingly, many of the foods that cause hypoglycemia or that interfere with the proper functioning of nitric oxide in the Nomad also inhibit your metabolism and cause weight gain.
- **Irritate the gut.** Many foods contain ingredients that can irritate the lining of the Nomad's gut, causing fatigue and inflammation. Many mold-containing foods and fungi can cause increased inflammation in Nomads.
- **Contain gluten.** Gluten is a protein found in many grains that can irritate the gut lining in sensitive individuals. If you are a Nomad and have discovered that you have white lines on your fingerprints, you'll want to limit your intake of gluten-containing foods.

Some foods on the *Nomad Don'ts* list need only be avoided for a short period of time so that you can regain your balance. After 3–6 months, you can reintroduce them back into your diet in modest amounts. These foods are identified by a black dot (•) icon. Of course, if you are battling an illness or feel your weight beginning to creep back up, you may want to ramp up your compliance by avoiding these foods again for a while.

## Unlisted Foods

Unlisted foods are foods that don't appear to do much good or bad. They are essentially neutral, and can be used judiciously (2–5 times weekly). The nutrients in them will benefit you, but they won't specifically help you restore balance to your genes or health to your cells. Feel free to eat them—but don't neglect the foods I recommend. The GenoType Diet is always evolving and I'm frequently adding new foods, so check in with my Web site (www.genotypediet.com), especially if you have questions about a specific food.

Well, Nomad, now is the time to put word and intent into action. From here you'll proceed to Chapter 14 to learn how you can get the most from the GenoType Diets. After that, we're off to Chapter 20 and the Nomad food, supplement, and exercise prescriptions.

# PART IV

# The GenoType Diets

## Six Individual Roads to Health

# CHAPTER FOURTEEN

# Getting the Most from Your GenoType Diet

Although many readers will be familiar with the basics of good nutrition and how to be an educated and informed food shopper, it's still a good idea to review some of the fundamentals of healthy eating. Here are some of the basic rules of healthy living that I share with all my patients regardless of their GenoType.

## The Ten Commandments of Any Successful Lifestyle

**1. It's what you eat, not what you avoid, that moves you forward.** As naturopathic medical students, we were to look for the root causes of disease and to always begin with the diet. However, my instructors were very much into allergy testing and elimination-type diets, and often by the time we were done with an examination, the patient was left with a glass of lemon water and a rice cake! Twenty-five years in clinical practice later,

I now know different. Sure, sometimes if you tell patients what to avoid they'll get "less sick," but in reality it is what you tell people to eat that actually makes them healthy. So when you start your GenoType Diet, begin by emphasizing what your GenoType best thrives upon, then slowly move some of the nonrecommended food out of the pantry and refrigerator.

**2. Don't eat when nervous or stressed.** Eating when tense or under circumstances that prohibit you from relaxing can have major effects on your digestion.

**3. Don't eat a major meal after 7 P.M.** If you are interested in keeping lean and fit, you will want to move your main meal to earlier in the day. Studies show that between two groups of research subjects eating the same meals but at different times, the group that ate their main meal at night gained weight, whereas the group that ate their main meal in the afternoon did not.

**4. Don't exercise to exhaustion.** Take yourself to your maximum, but go no further. If you need to lie down after you exercise, you're going at it too intensely.

**5. Don't diet.** This may seem like an odd strategy in a diet book, but here's my point: Many thrifty Warriors and Gatherers are used to "going on" and "going off" diets. The GenoType Diet is a road map, not a straitjacket. If you gradually incorporate the GenoType Diet into your daily life, the benefits will be apparent. One clue: The right diet will always make you feel better, not worse.

**6. Get up when you wake up.** As soon as your eyes open, get out of bed and begin your day. This will help synchronize your sleep-awake cycle and also allow you to start your day at peak performance. Trying to catch a few extra winks actually will wind up making you feel worse.

**7. Never go to bed stressed.** Take a little time to de-stress before bedtime. Take a hot bath. Watch a comedy on DVD. Talk about things with your mate or a good friend.

**8. Try not to combine starches and proteins.** Good food combining (eating meats with high-fiber vegetables and carbohydrates by themselves) enhances the transit time of the meal through the gut, which lowers the demands on the immune cells that line the digestive tract. Decreasing the workload of the immune system in the gut increases its ability to function efficiently, which decreases inflammation. Don't stress over it, but if the dinner plate looks like good food combining, smile and pat yourself on the back.

**9. Express yourself.** Following your GenoType Diet should be an adventure, not an ordeal. Keep an open mind; explore new foods and new ways of preparing them. Share ideas with friends. Use tools like the Internet to community-build and find support and camaraderie.

**10. Take the good, leave the rest.** Every belief system inhibits growth. If something in your GenoType Diet doesn't seem to work for you, skip it for now and give yourself some time to accomplish what you can do. When you are comfortable with these initial changes, perhaps give the others a second try.

## Understanding the GenoType Diet Food Categories

OK, now let's take a look at each food category, the pros and cons for each GenoType, and how you can be a better shopper and consumer.

### Red Meats

Despite sources who continue to claim that all humans can be successful vegetarians, the anthropological record indicates that this is simply not true. Although not a wise choice of food for Teacher and Warrior Geno-Types, red meat from organic ranches and wild game are essential for the success of several GenoTypes, including the Hunter and Gatherer and, to a lesser degree, the Nomad and Explorer. However, fatty, steroid- and antibiotic-laced animals raised under inhumane conditions not only fail to mimic the original environmental conditions but also tend to increase concentrations of pro-inflammatory fats and toxins. Gatherer and Explorer GenoTypes do better on leaner, rangier cuts of meat with a lower fat content, while Hunter and Nomad GenoTypes don't really need to worry about fat content. "Free-range" and "organic" are good monikers to look for, but best of all is "grass-fed," since even free-range or organic meats can be fed exclusively corn and soybeans, neither of which is part of the traditional diet of ruminant animals.

## Poultry

Some GenoTypes, such as the Gatherer and the Nomad, do not do very well with some forms of poultry, such as chicken. Others, like the Teacher, need to keep their total intake rather low. Hunters and Explorers do well with most poultry, especially the flying birds, which for them make better poultry choices than the terrestrial varieties. Flyers have high levels of dark meat, which means more myoglobin, a muscle protein that is a premium raw material for their metabolic furnaces. As with red meat, when consuming a food so high up the food chain you will need to search for the cleanest sources: free-range, hormone- and antibiotic-free sources are a must.

## Eggs

To some degree or another, all GenoTypes can use commercially available eggs as a source of a complete and inexpensive protein. Many farms are working to increase the omega-3 content of their eggs by including 10–20 percent of flax in the hen's diet, which in turn results in these eggs being higher in omega-3 fatty acids than conventional eggs. Try to find eggs "high in DHA"—one of the major omega-3 fatty acids, which are essential for optimal nerve and immune system health.

## Fish and Seafood

Hunters and Explorers (GenoTypes with reactive worldviews) and Warriors (a GenoType with rather thick, viscous blood) do best with the oily
types of ocean fish, which are rich in omega-3 and omega-6 fatty acids.
These help modulate and can correct pro-inflammatory capacities, and
surprisingly can also assist with balancing HPA (hypothalamus-pituitary
axis), raising spirits, and enhancing ability to respond resourcefully to stress.
Oily fish have oils throughout the fillet and in the belly cavity around the
gut, rather than only in the liver like white fish. Nomad and Teacher
GenoTypes do best with white fish, which have proteins that help balance and heal the intestinal lining and minimize bacterial overgrowth.
Make sure that purchased salmon is not "farm-raised" but rather "wild-
caught." All GenoTypes, but most important Explorers, should keep alert
to any news about fish contamination; the picture is constantly changing,
so you would best be advised to contact your regional Fish and Wildlife
Services for the most up-to-date information.

## Dairy Products

A relatively recent addition to the human diet, dairy products are very
GenoType-specific. Many dairy products that might be a problem in one
GenoType are sometimes a solution in another. For example, because they
can overreact to mycotoxins, Explorer GenoTypes should avoid the "blue"
category of cheeses: Gorgonzola, Limburger, Stilton, Roquefort, which
are made by inoculating loosely pressed curds with *Penicillium roqueforti*
or *Penicillium glaucum* molds. On the other hand, some of these cheeses
are "low-overgrowth" cheeses and actually can help rebuild the digestive
tract in Teacher GenoTypes. The softer cheeses, such as ricotta and moz-
zarella, are often well tolerated by Warrior and Nomad GenoTypes,
whereas the Hunter GenoType does poorly on most, if not all, cheeses.

## Vegetable Proteins

An author once wrote that if we were to pile up each and every food that humans have eaten since the beginning of our species, the acorn pile would probably still be the tallest. Often resorted to in historical times of scarcity, vegetable proteins, typically in the form of beans, nuts, and seeds, are a good protein source for Hunters—if they are chosen well. Many of the nuts and seeds in the human diet are GenoType-specific, especially those that contain allergens and lectins. Try to find vendors with high turnover, so you can be sure of getting your seeds, nuts, and legumes in as fresh a state as possible.

Unlike animal proteins, many vegetable proteins are not "complete proteins" in that they do not contain all of the essential amino acids; a protein missing one or more of these is an "incomplete protein." Most incomplete proteins can be combined to make a complete spectrum of all the essential amino acids. In the old days, nutritionists thought that incomplete proteins had to be combined at the same meal. However, we now know that if you consume a wide range of vegetable proteins, it is virtually guaranteed that you will wind up with all the amino acids that you need. Soy is an especially good food for the Teacher and Warrior Geno-Types; it is rich in isoflavone antioxidants, which help maintain the integrity of DNA. It has other unique anticancer components as well. You may occasionally read negative things about soy, and in some GenoTypes they may well be true, but if you are Warrior or Teacher, it's a good food.

## Fats and Oils

Oil choices are often GenoType-specific, with perhaps the exception of olive oil. Reactive GenoTypes such as the Hunter and the Explorer do best on oils that are monounsaturated, or a combination of monounsaturated and saturated. Tolerant GenoTypes such as the Teacher and the Nomad do better with short-chain fatty acids, such as the butyrate in ghee (clarified butter), which are marvelous for rebuilding the digestive tract and enhancing proper assimilation. The proper choice of oils and fats is also essential to getting the thrifty metabolisms of the Warrior and Gatherer back into shape. Always try to buy high-quality oils, preferably cold-pressed when appropriate. Oils do go bad (rancidify), so make a point of never buying more than you can use within two months.

## Carbohydrates

No food category has experienced the change in fortune that carbohydrates have undergone in the public consciousness. From the low-fat, high-carb eighties to the no-carb nineties, we've seen this class of food at the extremes of every diet spectrum. Of course, the reality is that certain carbohydrates are good for certain people. If you are gluten-sensitive, like the Hunter and Nomad GenoTypes, you'll want to watch your intake of gluten- and lectin-containing foods. If you have a bacterial overgrowth problem like the Teacher GenoType, you'll want to emphasize those foods that produce only slight residue. If you are thrifty like the Warrior or Gatherer, you'll want to increase your use of low-glycemic carbohydrates. If you've discovered that you have fingerprint white lines and follow these carbohydrate recommendations, the lining of your intestines will improve within a few months. However, it can take about a year for the fingerprint white lines to disappear. Many types of joint and muscle problems, such as arthritis and fibromyalgia, are inflammatory conditions that are worsened by wheat in the diet. An early sign that you are choosing carbohydrates correctly for your GenoType will be the welcome disappearance of morning aches and pains caused by stiffness.

## Live Foods

Live foods are enzyme-rich plant foods such as vegetables, kelps, and mushrooms. For each GenoType, the best live food choices are designed to be low in lectins, allergens, chitinase, pesticides, known genetically modified species, and molds. The superb choices for each GenoType are high in fiber, lignans, isoflavones, antioxidants, and all sorts of other goodies specific for the metabolisms and health problems of that GenoType. Choose organic, pesticide-free, non-irradiated, non-genetically modified vegetables and wash, wash, wash.

## Fruits

The best fruits for each are rich in antioxidants, vitamins, and fiber. In particular, berries and cherries are super antioxidant foods. Many of these plant antioxidants are specific to certain tissues of the body, which make them especially important nutrients for GenoTypes with health issues in those areas. For example, many of the blue-pigmented fruits have antioxidants that can heal joint tissue, while the antioxidants in yellow-pigmented fruits protect the delicate tissues of the eye and ovaries. What is especially great about the wide choices of fruits is how well they can substitute for carbohydrate cravings that can result from the severe grain restrictions in the early stages of the program. All fruits should be washed with a mild soap and rinsed for at least one minute.

## Spices

Spices have an ancient relationship with humans and played a prominent role in early medical treatment. GenoType Hunter does best with spices that support immune-system well-being by minimizing inflammation and reducing stress. Warriors and Gatherers can enhance their metabolisms by increasing their use of thermogenic spices. Teachers can benefit from the antimicrobial actions of many spices, Explorers from their detoxification benefits.

## Beverages

All the GenoTypes can benefit from the gene-protecting polyphenols found in green tea. Like fruits and live foods, many juices are GenoType-specific. All GenoTypes should avoid drinks sweetened with high-fructose corn syrup and drinks with phosphoric acid, such as diet colas. Coffee can be beneficial for Teachers and Warriors, but should be used in moderation by Nomads and Gatherers and avoided by Hunters and Explorers.

## Condiments

Sweeteners and other condiments are usually specific to GenoType. Many commercial condiments contain additives and preservatives that should be avoided by all GenoTypes. Alternative versions of commercial condiments that are acceptable for your GenoType can often be made at home from allowable ingredients. Gatherers, Warriors, and Nomads will want to be on the lookout for high-fructose corn syrup and other hidden sugars. Many sauces and preparations contain thickeners and gums that should be avoided by the Hunter, Teacher, and Explorer GenoTypes.

# GenoType Supplements

GenoTypes vary in their need for particular vitamins, minerals, and herbs, and a specific supplement plan is detailed for each GenoType as part of its diet chapter. But here are a few commonsense points I always make to my patients:

**1. Buy the best quality that you can afford.** Vitamins, herbs, and supplements vary greatly in quality and in shelf life. SAMe, for example (a great supplement for enhancing methylation), literally loses its potency from the moment of its manufacture. Kept in a hot tractor-trailer on its way from the factory to the store and you've got virtually zero biologic activity. Check your herbals. Are their contents standardized for active ingredients, or did they just chop up some dried herb and put it into a capsule?

**2. Buy from a well-regarded manufacturer.** Do they have a good reputation in the industry? Can they be trusted to provide an assay of their formulas to assure potency? Do they check for high bacteria counts in their supplements? Health food stores, consumer magazines, and pharmacies are often good places for discussing the relative merits of one supplement manufacturer over another.

**3. Take only the supplements that you need.** These days my desktop is covered with bottles of supplements that patients bring to their consultations. Many times, individual nutrients show up again and again in different formulas. This redundancy is not only wasteful—it can be dangerous.

## GenoType Exercise

Each of the GenoTypes has its own exercise plan—for both physical and emotional well-being. In addition to providing the recognized fitness benefits, exercise can be a wonderful way to fight the biochemical effects of stress. Since each GenoType has its own unique stress profile, a unique exercise plan is detailed for each GenoType as part of its diet chapter. But here are a few commonsense points I try to make to my patients:

If you aren't accustomed to a regular exercise regimen, these tips will help you get started—and keep moving:

**1. Find an exercise buddy.** A brisk morning walk or a game of tennis with a friend can be a great motivator.

**2. Vary your routine.** If boredom sets in, change your activity every three days. Tired of exercising first thing every morning? Move your exercise time to the evening for a week, then see which of the two makes you feel better.

**3. Challenge yourself with specific goals.** Don't be afraid to push yourself a little. For many of us exercise will be a distinct change of lifestyle, and our bodies often resist change, sometimes even when the change is good for us. Set up a weekly exercise strategy—where are you now and where do you want to be? However, remember what I said above: Don't take yourself to the point of exhaustion.

Now that you've mastered the basics, turn the page and let's get to the specific diet, exercise, and supplement plan for your GenoType.

# CHAPTER FIFTEEN

# The Hunter Diet

**W**elcome, Hunter! This chapter contains all the information that you'll need to get started on the diet for your GenoType. Diet buddy support, new research, and help with recipes and meal planning are only a mouse click away at the official GenoType Diet Web site (www.genotypediet.com).

The Hunter Diet is broken down by food category. Each category (Red Meats, Poultry, etc.) contains two lists. The list on the left contains the *Hunter Superfoods,* foods that act as medicines in the Hunter body, balancing stress, regenerating genes, and repairing the digestive tract. Superfoods that help Hunters maintain their ideal weight, increase muscle mass, and decrease body fat are identified by a diamond (◊) icon. To gain the maximum benefit from the Hunter Diet, they should be routinely consumed.

The list on the right contains the *Hunter Toxins,* foods that Hunter GenoTypes would be wise to avoid. Some foods on the Hunter Toxin list need only be avoided for a short period of time so that you can regain your balance. After 3–6 months, you can reintroduce these foods back into your diet in modest amounts. They are identified by a black dot (●) icon. Of

course, if you are battling an illness or feel your weight beginning to creep back up, you may want to ramp up your compliance by avoiding these foods again for a while.

If a food is not listed, it is essentially *neutral,* meaning that the nutrients in it will benefit you but won't specifically help you restore balance to your genes or health to your cells. Feel free to eat these foods—but don't neglect the foods I recommend. A complete list of all the foods I've tested is available online (www.genotypediet.com).

## Red Meats

*Portion size: About the size of your hand (4–6 ounces)*
*Frequency: 3–5 times weekly*

| Superfoods to Emphasize | Toxins to Limit or Avoid |
| --- | --- |
| Beef ◊ | Bacon |
| Beef, bone soups and broths | Beef heart ● |
| Beef, liver | Boar |
| Beef, tongue | Ham |
| Buffalo, bison ◊ | Kangaroo |
| Goat | Opossum |
| Lamb ◊ | Pork |
| Mutton | |
| Venison | |

## Poultry

*Portion size: About the size of your hand (4–6 ounces)*
*Frequency: 2–4 times weekly*

| Superfoods to Emphasize | Toxins to Limit or Avoid |
| --- | --- |
| Chicken | Chicken liver |
| Cornish hen | Duck liver |
| Duck ◊ | Goose liver |
| Grouse | Quail |

| Superfoods to Emphasize | Toxins to Limit or Avoid |
|---|---|
| Pheasant | |
| Squab | |
| Turkey | |

## Fish and Seafood

*Portion size: About the size of your hand (4–6 ounces)*
*Frequency: At least 4 times weekly*

| Superfoods to Emphasize | Toxins to Limit or Avoid |
|---|---|
| Bass, striped | Abalone |
| Chub | Barracuda |
| Cod | Bass, blue gill |
| Haddock | Bass, sea, lake ● |
| Hake | Bluefish |
| Herring | Catfish |
| Mackerel, Atlantic | Conch |
| Ocean pout | Crab ● |
| Pacific flounder | Grouper ● |
| Pacific sole | Frog |
| Pilchard | Muskellunge |
| Pompano | Octopus |
| Salmon, Atlantic, wild ◊ | Pollock, Atlantic |
| Salmon, chinook ◊ | Salmon, Atlantic, farm-raised |
| Salmon, sockeye ◊ | Squid, calamari |
| Sardine ◊ | Swordfish |
| Scrod | Tilefish |
| Smelt | Toothfish, Chilean sea bass |
| Sturgeon | Turtle |
| Trout, rainbow, wild ◊ | |
| Trout, sea ◊ | |
| Trout, steelhead, wild ◊ | |

## Eggs and Roes

*Portion size: 1 egg*
*Frequency: 7–9 servings per week*

| Superfoods to Emphasize | Toxins to Limit or Avoid |
|---|---|
| Carp roe | Caviar ● |
| Egg white, chicken ◊ | Herring roe |
| Egg white, duck ◊ | Goose egg |
| Egg, whole, chicken ◊ | Quail egg |
| Sailfish roe | |
| Salmon roe | |

## Dairy

*Portion size: Milk: 6 ounces; cheese: 2–4 ounces*
*Frequency: Cheeses: 3 times weekly; butter or ghee as desired*

| Superfoods to Emphasize | Toxins to Limit or Avoid |
|---|---|
| Butter | American cheese |
| Ghee (clarified butter) ◊ | Blue cheese |
| Kefalotyri cheese | Brie cheese |
| Manchego cheese | Camembert cheese |
| Parmesan cheese | Casein |
| Pecorino cheese | Cheddar cheese |
| Romano cheese | Cheshire cheese |
| | Colby cheese |
| | Cottage cheese |
| | Cream cheese |
| | Edam cheese ● |
| | Emmenthal, "Swiss" cheese ● |
| | Farmer cheese ● |
| | Feta cheese ● |
| | Gorgonzola cheese |
| | Gouda cheese |
| | Gruyère cheese |
| | Half-and-half |

| Superfoods to Emphasize | Toxins to Limit or Avoid |
| --- | --- |
| | Havarti cheese ● |
| | Jarlsberg cheese ● |
| | Limburger cheese |
| | Milk, buttermilk |
| | Milk, cow, skim or 2% |
| | Milk, cow, whole |
| | Milk, goat |
| | Monterey Jack cheese |
| | Mozzarella cheese ● |
| | Muenster cheese |
| | Neufchâtel cheese |
| | Paneer cheese |
| | Port du Salut cheese ● |
| | Provolone cheese |
| | Quark cheese |
| | Ricotta cheese |
| | Romanian Urdă ● |
| | Roquefort cheese |
| | Sour cream |
| | Stilton cheese |
| | String cheese |
| | Whey protein powder |
| | Yogurt |

## Vegetable Proteins

*Portion size: Nuts, seeds: ½ cup; nut butters: 2 tablespoons*
*Frequency: 3–7 times weekly*

| Superfoods to Emphasize | Toxins to Limit or Avoid |
| --- | --- |
| Adzuki bean | Beechnut |
| Almond | Brazil nut |
| Almond butter | Cashew ● |
| Black bean | Cashew butter ● |
| Black-eyed pea | Chestnut, European |
| Broad bean, fava ◊ | Copper bean |

| Superfoods to Emphasize | Toxins to Limit or Avoid |
|---|---|
| Butternut | Filbert, hazelnut ● |
| Carob ◊ | Hickory nuts |
| Chestnut, Chinese | Kidney bean |
| Chia seed | Lentils, all types ● |
| Flaxseed, linseed ◊ | Lentils, sprouted ● |
| Garbanzo bean, chickpea | Litchi |
| Great northern bean | Lotus seeds ● |
| Green, string bean ◊ | Lupin seeds ● |
| Haricot bean | Mung bean ● |
| Hemp seed | Natto |
| Lima bean ◊ | Navy bean |
| Lima bean flour ◊ | Peanut butter |
| Macadamia | Peanut flour |
| Peas | Peanuts |
| Pecan | Pinto bean ● |
| Pine nut, pignolia | Pinto bean, sprouted ● |
| Pumpkin seed ◊ | Pistachio ● |
| Safflower seed | Poppy seed |
| Snap bean ◊ | Soybean bean |
| Sesame seed ◊ | Soybean meal |
| Sesame butter, tahini ◊ | Soybean pasta |
| Sesame flour ◊ | Soybean, tempeh |
| Walnut | Soybean, tofu |
| Watermelon seeds | Sunflower seed |
| Yeast, baker's | Tamarind bean |
|  | Yard-long bean ● |

## Fats and Oils

*Portion size: 1 tablespoon*
*Frequency: 3–9 times weekly*

| Superfoods to Emphasize | Toxins to Limit or Avoid |
|---|---|
| Butter | Avocado oil |
| Camelina oil | Canola oil |
| Chia seed oil | Coconut oil ● |

| Superfoods to Emphasize | Toxins to Limit or Avoid |
|---|---|
| Cod liver oil ◊ | Corn oil |
| Flaxseed, linseed oil | Cottonseed oil |
| Ghee (clarified butter) ◊ | Grape seed oil |
| Hemp seed oil ◊ | Hazelnut oil ● |
| Herring oil ◊ | Lard |
| Olive oil | Margarine |
| Perilla seed oil | Oat oil ● |
| Pumpkin seed oil | Palm oil |
| Quinoa oil | Peanut oil |
| Rice bran oil ◊ | Safflower oil |
| Salmon oil | Shea nut oil ● |
| Sesame oil | Soybean oil |
| Walnut oil ◊ | Sunflower oil |
| | Wheat germ oil |

## Carbohydrates

*Portion size: ½ cup grains, cereals, rice; ½ muffin; 1 slice bread*
*Frequency: 2–3 times daily*

| Superfoods to Emphasize | Toxins to Limit or Avoid |
|---|---|
| 100% artichoke flour/pasta | Amaranth ● |
| Buckwheat, kasha, soba | Barley |
| Flaxseed bread | Cornmeal, hominy, polenta ● |
| Fonio | Essene, Manna breads ● |
| Job's tears | Kudzu ● |
| Millet | Lentil flour, dhal, poppadom ● |
| Quinoa | Oats, oat bran, oat flour ● |
| Rice bran | Rye |
| Rice flour, brown | Rye flour |
| Rice, basmati | Soybean flour |
| Rice, brown | Wheat, 100% sprouted ● |
| Rice, wild | Wheat, bran, germ |
| Taro | Wheat, bulgur |
| Teff | Wheat, durum, semolina |
| | Wheat, emmer |

| Superfoods to Emphasize | Toxins to Limit or Avoid |
|---|---|
| | Wheat, gluten |
| | Wheat, kamut |
| | Wheat, spelt ● |
| | Wheat, white flour |
| | Wheat, whole grain |

## Live Foods

*Portion size: 1 cup*
*Frequency: At least 4–5 servings daily*

| Superfoods to Emphasize | Toxins to Limit or Avoid |
|---|---|
| Artichoke ◊ | Alfalfa sprout |
| Asparagus | Aloe vera |
| Broccoflower | Avocado ● |
| Broccoli | Asparagus pea ● |
| Broccoli, Chinese ◊ | Bamboo shoot ● |
| Broccoli rabe ◊ | Beet green ● |
| Chicory, chicory root ◊ | Borage ● |
| Chinese kale, kai-lan | Brussels sprout |
| Dandelion | Carrot ● |
| Escarole | Cassava |
| Fenugreek | Cauliflower ● |
| Ginger ◊ | Collard greens ● |
| Grape leaves ◊ | Corn, popcorn ● |
| Jerusalem artichoke | Cucumber |
| Jew's ear (pepeao) ◊ | Eggplant ● |
| Kale | Kanpyo ● |
| Mushroom, enoki | Leek ● |
| Mustard greens ◊ | Lettuce, green leaf, iceberg ● |
| Okra | Mushroom, brown, cremini ● |
| Onion, all types | Mushroom, oyster |
| Parsnip | Mushroom, portobello |
| Pepper, chili, jalapeño ◊ | Mushroom, shiitake |
| Pumpkin | Mushroom, straw ● |
| Rowal ◊ | Mushroom, white, silver dollar |

| Superfoods to Emphasize | Toxins to Limit or Avoid |
|---|---|
| Rutabaga | Olive, black |
| Sea vegetables, kelp ◊ | Olive, green ● |
| Sea vegetables, spirulina | Pickle, brine |
| Sea vegetables, wakame ◊ | Pickle, vinegar |
| Sweet potato | Pimento ● |
| Sweet potato leaves ◊ | Potato, white, red, with skin |
| Swiss chard | Purslane |
| Turnip | Quorn |
| Turnip greens ◊ | Rhubarb |
| | Sauerkraut ● |
| | Sea vegetables, agar |
| | Sea vegetables, Irish moss |
| | Spinach ● |
| | Tomatillo ● |
| | Tomato ● |
| | Water chestnut, matai |

## Fruits

*Portion size: 1 cup fruit or 1 medium-sized fruit*
*Frequency: At least 3 servings daily*

| Superfoods to Emphasize | Toxins to Limit or Avoid |
|---|---|
| Acai berry | Apple ● |
| Banana | Apricot ● |
| Blueberry | Asian pear |
| Canistel ◊ | Bitter melon |
| Cranberry ◊ | Blackberry |
| Crenshaw melon | Cantaloupe melon |
| Date ◊ | Cherry ● |
| Dewberry | Coconut meat ● |
| Elderberry | Durian |
| Goji, wolfberry | Fig ● |
| Gooseberry | Grape ● |
| Groundcherry | Honeydew |
| Grapefruit ◊ | Kiwi |

| Superfoods to Emphasize | Toxins to Limit or Avoid |
|---|---|
| Guava | Nectarine ● |
| Huckleberry | Orange |
| Jackfruit | Papaya |
| Lemon | Plantain |
| Lime | Plum ● |
| Lingonberry ◊ | Pomegranate ● |
| Loganberry | Prune ● |
| Mamey sapote | Raisin ● |
| Mango ◊ | Tamarillo |
| Passion fruit ◊ | Tangerine |
| Pawpaw | Strawberry |
| Peach | |
| Pear ◊ | |
| Pineapple ◊ | |
| Quince | |
| Sago palm | |
| Watermelon ◊ | |

## Spices

*Portion size: 1 teaspoon*
*Frequency: At least 1–2 servings daily*

| Superfoods to Emphasize | Toxins to Limit or Avoid |
|---|---|
| Anise | Acacia (gum arabic) |
| Chocolate ◊ | Caper |
| Cilantro ◊ | Caramel |
| Cinnamon ◊ | Chives ● |
| Clove | Cornstarch ● |
| Curry ◊ | Guarana |
| Dulse | Mace |
| Garlic | Nutmeg |
| Rosemary | Parsley ● |
| Saffron | Pepper, black |
| Sage | |
| Tarragon | |
| Turmeric ◊ | |

## Beverages

*Portion size: 6–8-ounces*
*Frequency: 2–4 servings daily*

| Superfoods to Emphasize | Toxins to Limit or Avoid |
|---|---|
| Cranberry juice | Apple juice ● |
| Grape juice | Beer |
| Grapefruit juice | Beet juice ● |
| Pear juice | Blackberry juice |
| Pineapple juice | Carrot juice ● |
| Tea, chamomile ◊ | Coffee ● |
| Tea, gingerroot | Cola beverages |
| Tea, green, kukicha, bancha ◊ | Cucumber juice |
| Tea, lemon balm ◊ | Liquor, distilled |
| Tea, rooibos | Milk, almond ● |
| Tea, yerba maté ◊ | Milk, coconut ● |
| | Milk, rice ● |
| | Milk, soy |
| | Orange juice |
| | Tangerine juice |
| | Tea, black |
| | Tea, kombucha |
| | Wine, red ● |
| | Wine, white ● |

## Condiments and Additives

*Portion size: 1 teaspoon*
*Frequency: Use as needed*

| Superfoods to Emphasize | Toxins to Limit or Avoid |
|---|---|
| Agave syrup | Aspartame |
| Arrowroot | BHA, BHT |
| Locust bean gum | Carrageenan |
| Molasses | Dextrose |
| Molasses, blackstrap | Fructose |
| Mustard powder | High-fructose corn syrup |

| Superfoods to Emphasize | Toxins to Limit or Avoid |
|---|---|
| Umeboshi plum | Ketchup |
| Vegetable glycerine | Konjac |
| Yeast, nutritional ◊ | Guar gum |
| Yeast extract spread, Marmite | Mastic gum ● |
| | MSG |
| | Maple syrup ● |
| | Mayonnaise |
| | Mayonnaise, tofu |
| | Miso |
| | Mustard, with vinegar |
| | Phytic acid |
| | Pickle relish |
| | Potassium bisulfite |
| | Potassium metabisulfite |
| | Sodium bisulfite |
| | Sodium metabisulfite |
| | Soy sauce, tamari, wheat-free |
| | Sugar, brown, white |
| | Vinegar, all types |
| | Worcestershire sauce |

# GenoType Hunter Supplement Guide

These supplements can help improve your results while on the Hunter Diet. Most are readily available in health food stores, while a few are a bit more obscure and can be difficult to find. Note that these supplements are solely recommended for use as part of the Hunter GenoType Diet program. To examine how other supplements you may be using rate with the GenoType Hunter, or for more detailed information on the science behind the use of this supplement protocol, visit the official GenoType Diet Web site at www.genotypediet.com. Remember to always discuss the use of any nutritional supplements with your physician before embarking on a supplement program.

*Supplements that tone down the Hunter's "inflammatory genes":*
- *Scutellaria baicalensis* (Chinese skullcap). Typical daily dose: 250–500 mg
- Fish oils. Typical daily dose: 750–1,000 mg
- Butyric acid. Typical daily dose: 750–1,000 mg

*Supplements that tone down the Hunter's "rapid aging genes" and help regulate the metabolism:*
- Curcumin. Typical daily dose: 100–500 mg
- Green tea. Typical daily dose: 2–3 cups of tea
- Folic acid. Typical daily dose: 800 mcg
- Chromium polynicotinate. Typical daily dose: 50–200 mcg

*Supplements that help Hunters get their stress levels under control:*
- Holy basil (*Ocimum sanctum* Linn). Typical daily dose: 200–400 mg
- Pantothenic acid (vitamin $B_5$). Typical daily dose: 250–500 mg
- *Rhodiola rosea* (roseroot). Typical daily dose: 250–500 mg

## GenoType Hunter Lifestyle

### Meal Planning

The hunter has a wide range of food combinations and cuisines available for imaginative meal planning. We've collected a variety of meal plans (even plans for families with different member GenoTypes) and hundreds of tasty recipes for each GenoType online at www.genotypediet.com.

### Exercise Guide

Hunter GenoTypes need regular, vigorous exercise to stay fit and healthy, reduce stress, and increase endurance. To accomplish this, you'll need about 40 minutes of exercise daily. Be sure to warm up for at least 5–10 minutes with some gentle stretching before beginning any aerobic activity.

*Less Demanding*
- Hiking
- Pilates or other forms of core strengthening
- Vigorous walking: level ground, at least three miles
- Moderate competitive sports (tennis, racquetball, volleyball)
- Light upper- and lower-body weight work

*More Demanding*
- Intense competitive sports (martial arts, basketball, soccer)
- Dancing
- Gymnastics
- Moderate resistance training
- Running

*Practice De-stressing*
A great technique to help Hunters change their reaction pattern to stress is the old Yogic technique "alternate nostril breathing." Although it sounds a little weird, it has actually been researched and shown to help balance the parasympathetic (restorative) and sympathetic (flight or flight) parts of the nervous system.

1. Close the right nostril with your right thumb and take a deep, long inhale through your left nostril.
2. Now close the left nostril with your right ring finger and little finger, and at the same time remove your thumb from the right nostril and take a deep, long exhale through the right nostril.
3. Do this five times.
4. Now reverse it.
5. Close the left nostril with your left thumb and take a deep, long inhale through your right nostril.
6. Now close the right nostril with your left ring finger and little finger, and at the same time remove your thumb from the left nostril and take a deep, long exhale through the left nostril.
7. This is one round. Start by doing three rounds, adding one per week until you are doing seven rounds.

# CHAPTER SIXTEEN

# The Gatherer Diet

**Welcome, Gatherer!** This chapter contains all the information you'll need to get started on the diet for your GenoType. Diet buddy support, new research, and help with recipes and meal planning are only a mouse click away at the official GenoType Diet Web site (www.genotypediet.com).

The Gatherer Diet is broken down by food category. Each category (Red Meats, Poultry, etc.) contains two lists. The list on the left contains the *Gatherer Superfoods,* foods that act as medicines, balancing stress, regenerating genes, and repairing the digestive tract. Superfoods that help the Gatherer maintain ideal weight, increase muscle mass, and decrease body fat are identified by a diamond (◊) icon. To gain the maximum benefit from the Gatherer Diet, they should be routinely consumed.

The list on the right contains the *Gatherer Toxins,* foods that Gatherer GenoTypes would be wise to avoid. Some foods on the Gatherer Toxin list need only be avoided for a short period of time so that you can regain your balance. After 3–6 months, you can reintroduce these foods back into your diet in modest amounts. They are identified by a black dot (•) icon. Of course, if you are battling an illness or feel your weight beginning to creep back up, you may want to ramp up your compliance by avoiding these foods again for a while.

If a food is not listed, it is essentially *neutral*, meaning that the nutrients in it will benefit you but won't specifically help you restore balance to your genes or health to your cells. Feel free to eat these foods—but don't neglect the foods I recommend. A complete list of all the foods I've tested is available online (www.genotypediet.com).

## Red Meats

*Portion size: About the size of your hand (4–6 ounces)*
*Frequency: 3–5 times weekly*

| Superfoods to Emphasize | Toxins to Limit or Avoid |
| --- | --- |
| Buffalo, bison | Beef heart |
| Caribou | Beef liver |
| Goal ◊ | Boar |
| Lamb ◊ | Ham |
| Mutton ◊ | Pork |
| Rabbit | Pork, bacon |
|  | Sweetbreads |
|  | Venison ● |

## Poultry

*Portion size: About the size of your hand (4–6 ounces)*
*Frequency: 2–5 times weekly*

| Superfoods to Emphasize | Toxins to Limit or Avoid |
| --- | --- |
| Emu ◊ | Chicken ● |
| Ostrich ◊ | Chicken liver |
| Pheasant | Cornish hens ● |
| Squab | Duck |
| Turkey ◊ | Duck liver |
|  | Goose |
|  | Goose liver |
|  | Grouse |

| Superfoods to Emphasize | Toxins to Limit or Avoid |
|---|---|
| | Guinea hen ● |
| | Partridge ● |
| | Quail |

## Fish and Seafood

*Portion size: About the size of your hand (4–6 ounces)*
*Frequency: At least 4 times weekly*

| Superfoods to Emphasize | Toxins to Limit or Avoid |
|---|---|
| Bullhead | Abalone |
| Butterfish, sablefish ◊ | Anchovy ● |
| Carp | Barracuda |
| Catfish | Bass, blue gill |
| Chub ◊ | Bass, sea, lake |
| Flounder | Bass, striped (farm-raised) ● |
| Gray sole ◊ | Bluefish |
| Grouper | Clam ● |
| Haddock | Conch |
| Hake | Crab |
| Halibut | Frog |
| Harvest fish ◊ | Lobster ● |
| Herring ◊ | Mackerel, Atlantic |
| Mahimahi ◊ | Mackerel, Spanish ● |
| Mullet ◊ | Muskellunge |
| Ocean pout | Mussels |
| Opaleye fish | Octopus |
| Orange roughy | Oyster |
| Perch | Pollock, Atlantic |
| Perch, ocean | Scallop |
| Pickerel (walleye) | Sea bream |
| Pike | Shad |
| Pilchard | Shrimp |
| Salmon, chinook ◊ | Skate |
| Salmon, sockeye ◊ | Smelt |
| Sardine ◊ | Squid, calamari ● |
| Sucker | Trout, farm-raised |

| Superfoods to Emphasize | Toxins to Limit or Avoid |
|---|---|
| Tilapia | Trout, sea ● |
| Tilefish ◊ | Trout, rainbow, wild |
| Tuna, bluefin ◊ | Trout, steelhead, wild ● |
| Tuna, skipjack | Turtle |
| Tuna, yellowfin | Weakfish ● |
| Turbot, European | |
| Whitefish | |
| Whiting | |
| Wolffish, Atlantic | |
| Yellowtail | |

## Eggs and Roes

*Portion size: 1 egg*
*Frequency: 7–9 servings per week*

| Superfoods to Emphasize | Toxins to Limit or Avoid |
|---|---|
| Egg, whole, chicken | Carp roe |
| Egg white, chicken ◊ | Caviar |
| Egg white, duck | Egg, whole, duck |
| Herring roe | Goose egg |
| | Quail egg |
| | Sailfish roe |
| | Salmon roe |

## Dairy

*Portion size: Milk: 6 ounces; cheese: 2–4 ounce; ghee, butter: 1 tsp.*
*Frequency: Cheeses: 3 times weekly; ghee: 4 times weekly*

| Superfoods to Emphasize | Toxins to Limit or Avoid |
|---|---|
| Cottage cheese ◊ | American cheese |
| Farmer cheese ◊ | Blue cheese |
| Ghee (clarified butter) | Brie cheese |
| Paneer cheese ◊ | Camembert cheese |

| Superfoods to Emphasize | Toxins to Limit or Avoid |
|---|---|
| Quark cheese | Casein |
| Romanian Urdă | Cheddar cheese |
| Ricotta cheese ◊ | Cheshire cheese |
| | Colby cheese |
| | Cream cheese |
| | Edam cheese |
| | Emmenthal, "Swiss" cheese |
| | Feta cheese ● |
| | Gjetost ● |
| | Gorgonzola cheese |
| | Gouda cheese |
| | Gruyère cheese ● |
| | Half-and-half |
| | Havarti cheese |
| | Jarlsberg cheese |
| | Kefalotyri cheese ● |
| | Kefir |
| | Limburger cheese |
| | Manchego cheese ● |
| | Milk, buttermilk |
| | Milk, cow, skim or 2% |
| | Milk, cow, whole |
| | Milk, goat ● |
| | Monterey Jack cheese |
| | Mozzarella cheese ● |
| | Muenster cheese |
| | Neufchâtel cheese |
| | Parmesan cheese |
| | Pecorino cheese |
| | Port du Salut cheese ● |
| | Provolone cheese |
| | Romano cheese |
| | Roquefort cheese |
| | Sour cream |
| | Stilton cheese |
| | String cheese |
| | Whey protein powder ● |
| | Yogurt |

# Vegetable Proteins

*Portion size: Nuts, seeds: ½ cup; nut butters: 2 tablespoons*
*Frequency: 3–7 times weekly*

| Superfoods to Emphasize | Toxins to Limit or Avoid |
|---|---|
| Almond ◊ | Adzuki bean ● |
| Almond butter | Beechnut |
| Butter bean ◊ | Black bean ● |
| Butternut | Black-eyed pea |
| Cannellini bean | Brazil nut ● |
| Carob ◊ | Broad bean, fava |
| Chestnut, Chinese | Cashew ● |
| Chia seed, pinole ◊ | Cashew butter ● |
| Flaxseed, linseed | Chestnut, European |
| Great northern bean ◊ | Copper bean ● |
| Green, string bean | Fava bean |
| Hickory nuts | Filbert, hazelnut |
| Lupin seeds ◊ | Garbanzo bean, chickpea |
| Moth bean | Haricot bean |
| Peas ◊ | Hemp seed |
| Pecan ◊ | Kidney bean |
| Pumpkin seed ◊ | Lentils, all types ● |
| Walnut | Lentils, sprouted ● |
| Watermelon seeds ◊ | Litchi |
| White bean | Mung bean |
| Winged bean | Natto |
| Yeast, baker's ◊ | Navy bean |
| | Peanut butter |
| | Peanut flour |
| | Peanuts |
| | Pine nut, pignolia |
| | Pinto bean ● |
| | Pinto bean, sprouted ● |
| | Pistachio ● |
| | Poppy seed |
| | Safflower seed |
| | Sesame seed ● |
| | Sesame butter, tahini ● |

| Superfoods to Emphasize | Toxins to Limit or Avoid |
|---|---|
| | Sesame flour ● |
| | Soybean bean |
| | Soybean meal |
| | Soybean pasta |
| | Soybean, sprouted |
| | Soybean, tempeh |
| | Soybean, tofu |
| | Sunflower seed ● |
| | Tamarind bean |

## Fats and Oils

*Portion size: 1 tablespoon*
*Frequency: 3–6 times weekly*

| Superfoods to Emphasize | Toxins to Limit or Avoid |
|---|---|
| Almond oil ◊ | Avocado oil |
| Apricot kernel oil | Canola oil ● |
| Camelina oil ◊ | Coconut oil ● |
| Chia seed oil | Cod liver oil ● |
| Flaxseed, linseed oil | Corn oil |
| Grape seed oil ◊ | Cottonseed oil |
| Hazelnut nut oil | Lard |
| Hemp seed oil | Margarine |
| Macadamia oil ◊ | Palm oil |
| Oat oil | Peanut oil |
| Olive oil ◊ | Pumpkin seed oil ● |
| Perilla seed oil | Safflower oil |
| Quinoa oil | Sesame oil ● |
| Rice bran oil ◊ | Soybean oil |
| Walnut oil ◊ | Sunflower oil |
| | Wheat germ oil |

# Carbohydrates

*Portion size: ½ cup grains, cereals, rice; ½ muffin; 1 slice bread*
*Frequency: 2–3 times daily*

| Superfoods to Emphasize | Toxins to Limit or Avoid |
| --- | --- |
| 100% artichoke flour/pasta | Buckwheat, kasha, soba ● |
| Amaranth ◊ | Cornmeal, hominy, polenta |
| Barley | Lentil flour, dhal |
| Essene, Manna bread | Poppadom ● |
| Flaxseed bread ◊ | Rice, brown ● |
| Fonio ◊ | Rice, white ● |
| Job's tears | Rice flour, brown ● |
| Millet ◊ | Rice flour, white ● |
| Oat bran | Rice, wild ● |
| Oat flour | Rye ● |
| Quinoa ◊ | Rye flour ● |
| Rice bran ◊ | Sorghum |
| Rice, basmati | Soybean flour |
| Teff | Tapioca, manioc, cassava |
| | Wheat, bran, germ |
| | Wheat, bulgur |
| | Wheat, durum, semolina |
| | Wheat, puffed |
| | Wheat, white flour |
| | Wheat, whole-grain |
| | Wheat, whole-grain emmer |
| | Wheat, whole-grain kamut |
| | Wheat, whole-grain spelt ● |

## Live Foods

*Portion size: 1 cup*
*Frequency: At least 4–5 servings daily*

| Superfoods to Emphasize | Toxins to Limit or Avoid |
|---|---|
| Asparagus | Alfalfa sprouts |
| Asparagus pea ◊ | Aloe vera |
| Bamboo shoot | Avocado ● |
| Celery ◊ | Beet ● |
| Chicory root | Bok choy, pak choi |
| Dandelion | Broccoflower ● |
| Escarole | Broccoli ● |
| Fennel | Brussels sprout ● |
| Fenugreek ◊ | Cabbage ● |
| Fiddlehead fern | Carrot ● |
| Jute, potherb | Cauliflower |
| Kanpyo | Chayote, pipinella |
| Mushroom, commercial ◊ | Chinese kale, kai-lan ● |
| Mushroom, cremini ◊ | Corn, popcorn |
| Mushroom, enoki ◊ | Cucumber ● |
| Mushroom, maitake ◊ | Eggplant ● |
| Mushroom, oyster ◊ | Grape leaves ● |
| Mushroom, portobello ◊ | Hearts of palm |
| Okra ◊ | Kale ● |
| Onion, all types | Leek |
| Pepper, bell | Mushroom, shiitake |
| Pepper, chili, jalapeño | Olive, black |
| Pimento | Olive, green |
| Sea vegetables, kelp ◊ | Oyster plant, salsify ● |
| Sea vegetables, spirulina ◊ | Parsnip ● |
| Shallots | Pickle, brine |
| Spinach | Pickle, vinegar |
| Sweet potato leaves | Potato, white with skin |
| Taro | Quorn |
| Tomato ◊ | Radish ● |
| Turnip greens | Radish sprouts ● |
| Water chestnut, matai | Rhubarb |
| Yam | Sauerkraut |

| Superfoods to Emphasize | Toxins to Limit or Avoid |
|---|---|
| Zucchini | Sea vegetables, agar |
| | Squash ● |
| | Sweet potato ● |
| | Watercress ● |

## Fruits

*Portion size: 1 cup fruit or 1 medium-sized fruit*
*Frequency: At least 3 servings daily*

| Superfoods to Emphasize | Toxins to Limit or Avoid |
|---|---|
| Apricot | Apple ● |
| Breadfruit ◊ | Asian pear |
| Carissa, natal plum | Banana ● |
| Cherimoya | Bitter melon |
| Cloudberry | Blackberry |
| Cranberry | Blueberry ● |
| Currants | Cantaloupe melon ● |
| Dewberry | Cherry ● |
| Elderberry ◊ | Coconut meat |
| Feijoa | Date |
| Gooseberry | Durian |
| Grapefruit ◊ | Grape ● |
| Guava ◊ | Honeydew ● |
| Lemon | Huckleberry |
| Lime | Kiwi |
| Lingonberry ◊ | Kumquat ● |
| Loganberry ◊ | Loquat |
| Mamey sapote | Muskmelon |
| Mulberry | Orange ● |
| Nectarine | Pear ● |
| Noni | Persian melon |
| Papaya | Persimmon |
| Peach ◊ | Plantain ● |
| Pineapple ◊ | Plum |
| Pummelo ◊ | Pomegranate |

| Superfoods to Emphasize | Toxins to Limit or Avoid |
|---|---|
| Raspberry | Prickly pear |
| Watermelon ◊ | Prune ● |
| | Raisin ● |
| | Rowanberry |
| | Spanish melon |
| | Star fruit, carambola |
| | Strawberry ● |
| | Tangerine ● |

## Spices

*Portion size: 1 teaspoon*
*Frequency: At least 1–2 servings daily*

| Superfoods to Emphasize | Toxins to Limit or Avoid |
|---|---|
| Basil | Acacia (gum arabic) |
| Cardamom | Allspice ● |
| Chocolate ◊ | Anise ● |
| Cilantro ◊ | Caper |
| Cinnamon ◊ | Chili powder |
| Curry ◊ | Clove ● |
| Fennel | Coriander ● |
| Garlic | Cornstarch |
| Paprika | Guarana |
| Parsley ◊ | Mace |
| Turmeric ◊ | Pepper, black |
| Saffron | Rosemary ● |
| | Vanilla ● |

## Beverages

*Portion size: 6–8-ounces*
*Frequency 2–4 servings daily*

| Superfoods to Emphasize | Toxins to Limit or Avoid |
|---|---|
| Cranberry juice | Apple juice |
| Elderberry juice | Beer |
| Kombucha tea | Blackberry juice |
| Lemon and water | Blueberry juice |
| Pummelo juice ◊ | Cabbage juice |
| Milk, rice | Cherry juice ● |
| Noni juice | Coffee ● |
| Tea, black | Cola beverages |
| Tea, chamomile | Cucumber juice |
| Tea, green, kukicha ◊ | Liquor, distilled |
| Tea, gingerroot | Milk, almond ● |
| Tea, ginseng ◊ | Milk, coconut |
| Tea, lemon balm | Milk, soy |
| Tea, yerba maté ◊ | Orange juice |
| Watermelon juice | Pear juice |
| | Pomegranate juice |
| | Tangerine juice |
| | Tomato juice |
| | Wine, red |
| | Wine, white ● |

## Condiments and Additives

*Portion size: 1 teaspoon*
*Frequency: Use as needed*

| Superfoods to Emphasize | Toxins to Limit or Avoid |
|---|---|
| Epazote | Aspartame |
| Fruit pectin ◊ | Barley malt |
| Kelp powder | Carrageenan |
| Locust bean gum | Caramel, caramel coloring |

| Superfoods to Emphasize | Toxins to Limit or Avoid |
|---|---|
| Molasses | Dextrose |
| Molasses, blackstrap | Ethyl maltol |
| Reihan, Tulsi (Holy basil) | Fructose |
| Sea salt | Guanosine monophosphate |
| Stevia | Guar gum |
| Umeboshi plum, vinegar | High-fructose corn syrup |
| Vegetable glycerine | Honey ● |
| Yeast spread, Marmite | Ketchup |
| Yeast, nutritional | Invertase |
| | Inverted sugar syrup |
| | Maltitol |
| | Maple syrup ● |
| | Mayonnaise |
| | Mayonnaise, tofu |
| | Miso |
| | Pickle relish |
| | Polysorbate |
| | Potassium sorbate |
| | Rice syrup |
| | Sorbic acid |
| | Sorbitol |
| | Soybean sauce, tamari, wheat-free |
| | Sucralose |
| | Sugar, brown, white |
| | Worcestershire sauce |

## GenoType Gatherer Supplement Guide

These supplements can help improve your results while on the Gatherer Diet. Most are readily available in health food stores, while a few are a bit more obscure. Note that these supplements are solely recommended for use as part of the Gatherer GenoType Diet program. To examine how other supplements you may be using rate with GenoType Gatherer, or for more detailed information on the science behind the use of this supplement protocol, visit the official GenoType Diet Web site at www.genotypediet.com.

Remember to always discuss the use of any nutritional supplements with your physician before embarking on a supplement program.

*Supplements that tone down the Gatherer's "AGE glycation genes":*
- Quercetin. Typical daily dose: 250–500 mg
- Carnosine.* Typical daily dose: 50–100 mg
- Yerba maté (*Ilex paraguariensis*). Typical daily dose: tea, 2 cups; supplement, 200–400 mg

*Supplements that enhance the Gatherer's mood, increase the efficiency of the hormonal response, and help regulate the metabolism and appetite-control center:*
- *Magnolia officinalis.* Typical daily dose: 250–400 mg
- Melatonin or methylcobalamine. Typical daily dose: melatonin, 3 mg before bed; methylcobalamine, 50 mcg before bed
- Coleus forskolin. Typical daily dose: Standardized root extract, 150–250 mg

*Supplements that help turn down the Gatherer's thrifty genetic tendencies:*
- Salacia oblonga. Typical daily dose: 250–400 mg
- Resveratrol. Typical daily dose: 50–100 mg
- Lipoic acid. Typical daily dose: 50–100 mg
- Green tea. Typical daily dose: 2–3 cups of tea

# GenoType Gatherer Lifestyle
## Meal Planning

The Gatherer has a wide range of food combinations and cuisines available for imaginative meal planning. We've collected a variety of meal plans (even plans for families with different member GenoTypes) and hundreds of tasty recipes for each GenoType online at www.genotypediet.com.

---

* *Note:* Do not confuse this supplement with the far more available amino acid l-carnitine.

## Exercise Guide

The Gatherer exercise protocol is designed to increase muscle growth, clear out toxins that are stored in body-fat tissue, and enhance insulin efficiency. Gatherer GenoTypes do best with stretching and lengthening exercises that also have a resistance component to them. Try to exercise for 30–40 minutes at least 6 days per week. Be sure to do warm-up stretching for 5 minutes before you begin, and cool down with stretching for 5 minutes after you've finished. Depending on your current level of conditioning, choose from the following list of exercises:

*Less Demanding*
- Golf: nine holes, no golf cart
- Vigorous walking: level ground, at least two miles
- Active hatha yoga
- Swimming
- Hiking: moderate pace, some hilly terrain
- Pilates

*More Demanding*
- Resistance training: light circuit training, or light hand weights (2–5 pounds) used with walking or hiking
- Tennis
- Aerobics
- Martial arts or competitive sports

*Soothe and Detoxify*
Take a detoxifying sauna (preferably one of the new "far-infrared" types) at least twice weekly when beginning the program. The sauna will help remove toxic chemicals from your fat tissues and mobilize the removal of stored fat. Far infrared is claimed to penetrate the body's tissues to a depth of 1 to 3 inches, far deeper than steam or other types of dry heat. This "deep heating," along with sweating, is thought to be responsible for the health benefits associated with these infrared rays.

## How to Do It

- Always consult with your personal physician before using any sauna. People with multiple sclerosis, lupus, adrenal suppression, hyperthyroidism, or hemophilia should not use a sauna. Pregnant women, children under five years of age, and people with artificial joints, metal pins, or silicon breast implants should not use a sauna.

- If you do not feel well at any point during your sauna, leave immediately, especially if you feel weak, light-headed, dizzy, or nauseated. If symptoms do not resolve, seek medical attention.

- The best time for a sauna is first thing in the morning or last thing before bed.

- Drink one glass of water for every 15 minutes you are in the sauna.

- Start with 15 minutes per day and gradually increase. After a few weeks, increase to 30–40 minutes. If you suffer from serious medical problems, start with 10 minutes.

- The number of sessions and amount of time spent in the sauna should be discussed with your physician. This should be individualized. Typically, 2–3 times weekly is fine.

- After the sauna session, rest and cool down for at least 15 minutes with a shower or favorite relaxation technique.

# CHAPTER SEVENTEEN

# The Teacher Diet

**W**elcome, Teacher! This chapter contains all the information that you'll need to get started on the diet for your GenoType. Diet buddy support, new research, and help with recipes and meal planning are only a mouse click away, at the official GenoType Diet Web site (www.genotypediet.com).

The Teacher Diet is broken down by food category. Each category (Red Meats, Poultry, etc.) contains two lists. The list on the left contains the *Teacher Superfoods,* which are foods that in the Teacher body act as medicines, balancing stress, regenerating genes, and repairing the digestive tract. Superfoods that help Teacher GenoTypes maintain ideal weight, increase muscle mass, and decrease body fat are identified by a diamond (◊) icon. To gain the maximum benefit from the Teacher Diet, they should be routinely consumed.

The list on the right contains the *Teacher Toxins,* foods that Teacher GenoTypes would be wise to avoid. Some foods on the Teacher Toxin list need only be avoided for a short period of time so that you can regain your balance. After 3–6 months, you can reintroduce these foods back into your diet in modest amounts. They are identified by a black dot (●) icon. Of

course, if you are battling an illness or feel your weight beginning to creep back up, you may want to ramp up your compliance by avoiding these foods again for a while.

If a food is not listed, it is essentially *neutral*, meaning that the nutrients in it will benefit you but won't specifically help you restore balance to your genes or health to your cells. Feel free to eat these foods—but don't neglect the foods I recommend. A complete list of all the foods I've tested is available online (www.genotypediet.com).

## Red Meats

*Portion size: About the size of your hand (4–6 ounces)*
*Frequency: 0–2 times weekly*

| Superfoods to Emphasize | Toxins to Limit or Avoid |
|---|---|
| Goat | Beef |
| Mutton | Beef heart |
|  | Beef liver |
|  | Beef tongue |
|  | Boar |
|  | Bone soup |
|  | Buffalo, bison |
|  | Caribou ● |
|  | Ham |
|  | Horse |
|  | Kangaroo |
|  | Lamb ● |
|  | Marrow soup |
|  | Moose |
|  | Opossum |
|  | Pork |
|  | Pork, bacon |
|  | Rabbit |
|  | Sweetbreads |
|  | Veal |
|  | Venison |

# Poultry

*Portion size: About the size of your hand (4–6 ounces)*
*Frequency: 1–3 times weekly*

| Superfoods to Emphasize | Toxins to Limit or Avoid |
|---|---|
| Emu ◊ | Chicken ● |
| Ostrich ◊ | Chicken liver ● |
| Squab | Cornish hens ● |
| Turkey ◊ | Duck |
| | Duck liver |
| | Goose |
| | Goose liver |
| | Grouse |
| | Guinea hen |
| | Partridge |
| | Pheasant |
| | Quail |

# Fish and Seafood

*Portion size: About the size of your hand (4–6 ounces)*
*Frequency: 3–4 times weekly*

| Superfoods to Emphasize | Toxins to Limit or Avoid |
|---|---|
| Abalone | Anchovy |
| Bullhead ◊ | Barracuda |
| Butterfish | Bass, blue gill |
| Carp | Bass, sea |
| Chub ◊ | Bass, striped |
| Cod ◊ | Bluefish |
| Croaker | Catfish |
| Cusk | Clam |
| Drum | Conch |
| Halfmoon fish | Crab |
| Mahimahi | Eel |
| Monkfish | Flounder |

| Superfoods to Emphasize | Toxins to Limit or Avoid |
|---|---|
| Mullet ◊ | Frog |
| Muskellunge ◊ | Gray sole |
| Ocean pout | Grouper |
| Parrotfish | Haddock |
| Perch ◊ | Hake |
| Pickerel (walleye) ◊ | Halibut |
| Pike ◊ | Harvest fish |
| Pilchard | Herring ● |
| Pollock, Atlantic | Jellyfish, dried, salted |
| Pompano ◊ | Lobster |
| Porgy | Mackerel, Atlantic |
| Rosefish | Mackerel, Spanish ● |
| Salmon, Alaskan ◊ | Mussels |
| Salmon, chinook ◊ | Octopus |
| Salmon, sockeye ◊ | Opaleye fish |
| Sardine ◊ | Orange roughy |
| Scrod ◊ | Oyster |
| Sea bream | Salmon, farmed |
| Smelt | Scallop |
| Snail, escargot ◊ | Scup |
| Sucker | Shad |
| Sunfish, pumpkinseed | Shark |
| Tilapia | Sheepshead fish |
| Tuna, skipjack | Shrimp |
| Tuna, yellowfin | Skate |
| Turbot, European | Sole |
| Whitefish ◊ | Squid, calamari |
| | Swordfish |
| | Tilefish |
| | Trout, farmed |
| | Trout, sea |
| | Trout, rainbow, wild ● |
| | Trout, steelhead, wild ● |
| | Turtle |
| | Weakfish |
| | Whiting ● |
| | Wolffish, Atlantic |
| | Yellowtail ● |

## Eggs and Roes

*Portion size: 1 egg*
*Frequency: 6–9 servings per week*

| oods to Emphasize | Toxins to Limit or Avoid |
|---|---|
| Egg white, chicken ◊ | Carp roe ● |
| Egg, whole, chicken ◊ | Caviar |
| Egg yolk, chicken | Goose egg |
| Sailfish roe ◊ | Egg white, duck |
| Salmon roe ◊ | Egg, whole, duck |
|  | Herring roe |
|  | Quail egg |

## Dairy

*Portion size: Milk: 6 ounces; cheese: 2–4 ounces; ghee: as desired*
*Frequency: 4–6 times weekly*

| oods to Emphasize | Toxins to Limit or Avoid |
|---|---|
| Brie cheese | American cheese |
| Colby cheese | Butter ● |
| Edam cheese | Casein |
| Emmenthal, "Swiss" cheese | Cheddar cheese ● |
| Ghee (clarified butter) | Cheshire cheese |
| Gorgonzola cheese ◊ | Cottage cheese |
| Gouda cheese ◊ | Cream cheese |
| Gruyère cheese | Farmer cheese |
| Havarti cheese ◊ | Feta cheese ● |
| Jarlsberg cheese | Half-and-half |
| Manchego cheese ◊ | Kefalotyri cheese |
| Milk, buttermilk | Kefir |
| Monterey Jack cheese | Limburger cheese |
| Muenster cheese ◊ | Milk, buffalo ● |
| Parmesan cheese ◊ | Milk, cow, skim or 2% |
| Pecorino cheese ◊ | Milk, cow, whole |
| Provolone cheese ◊ | Milk, goat |
| Romano cheese ◊ | Mozzarella cheese |

| Superfoods to Emphasize | Toxins to Limit or Avoid |
|---|---|
| Stilton cheese | Neufchâtel cheese |
| Yogurt | Paneer cheese |
| | Port du Salut cheese ● |
| | Quark cheese |
| | Ricotta cheese |
| | Romanian Urdă |
| | Sour cream |
| | String cheese |
| | Whey protein powder |

## Vegetable Proteins

*Portion size: Nuts, seeds: ½ cup; nut butters: 2 tablespoons*
*Frequency: 7–10 times weekly*

| Superfoods to Emphasize | Toxins to Limit or Avoid |
|---|---|
| Adzuki bean | Black-eyed pea |
| Almond ◊ | Brazil nut ● |
| Almond butter ◊ | Cashew ● |
| Black bean | Cashew butter |
| Broad bean, fava ◊ | Copper bean |
| Cannellini bean | Filbert, hazelnut ● |
| Carob ◊ | Garbanzo bean, chickpea ● |
| Chestnut, European | Kidney bean ● |
| Flaxseed, linseed ◊ | Lima bean |
| Great northern bean | Litchi |
| Green, string bean | Lotus root |
| Haricot beans | Lotus seeds |
| Hemp seed | Lupin seeds |
| Lentils, all types | Macadamia ● |
| Natto ◊ | Mung bean |
| Peanut butter ◊ | Navy bean |
| Peanut flour | Pistachio |
| Peanuts ◊ | Poppy seed ● |
| Peas ◊ | Pumpkin seed ● |
| Pecan ◊ | Sapodilla |

| Superfoods to Emphasize | Toxins to Limit or Avoid |
|---|---|
| Pine nut, pignolia | Sesame butter, tahini ● |
| Pinto bean ◊ | Sesame flour ● |
| Safflower seed ◊ | Sesame seed ● |
| Soybean | Sunflower seed ● |
| Soybean pasta | Tamarind bean |
| Soybean, sprouted | Winged beans |
| Soybean, tempeh | |
| Soybean, tofu | |
| Walnut ◊ | |
| Watermelon seeds | |
| Yeast, baker's ◊ | |

## Fats and Oils

*Portion size: 1 tablespoon*
*Frequency: 3–6 times weekly*

| Superfoods to Emphasize | Toxins to Limit or Avoid |
|---|---|
| Almond oil | Apricot kernel oil |
| Black currant seed oil | Avocado oil ● |
| Borage seed oil | Butter ● |
| Camelina oil ◊ | Canola oil |
| Chia seed oil | Corn oil |
| Coconut oil ◊ | Cottonseed oil |
| Flaxseed, linseed oil ◊ | Hazelnut nut oil ● |
| Ghee (clarified butter) | Lard |
| Grape seed oil ◊ | Margarine |
| Hemp seed oil ◊ | Palm oil |
| Herring oil | Peanut oil ● |
| Oat oil | Sesame oil ● |
| Olive oil ◊ | Soybean oil ● |
| Perilla seed oil ◊ | Sunflower oil ● |
| Quinoa oil ◊ | Walnut oil ● |
| Rice bran oil ◊ | |
| Safflower oil | |
| Salmon oil | |

## Carbohydrates

*Portion size: ½ cup grains, cereals, rice; ½ muffin; 1 slice bread*
*Frequency: 2–5 times daily*

| Superfoods to Emphasize | Toxins to Limit or Avoid |
| --- | --- |
| 100% artichoke pasta ◊ | Amaranth ● |
| Buckwheat, kasha, soba | Barley |
| Flaxseed bread ◊ | Corn, cornmeal, polenta ● |
| Kudzu ◊ | Essene, Manna bread ● |
| Lentil flour, dhal, poppadom | Fonio ● |
| Oat bran | Job's tears ● |
| Oat flour | Millet ● |
| Quinoa ◊ | Rice, basmati ● |
| Rice bran | Rice, white ● |
| Rice flour, brown | Rice flour, white ● |
| Rice, brown | Rye |
| Rice, wild | Rye flour |
| Sorghum | Tapioca, manioc, cassava |
| Taro | Wheat, durum ● |
| Teff | Wheat, white flour ● |
| Wheat, emmer, einkorn | Wheat, 100% sprouted ● |
| Wheat, kamut | |
| Wheat, spelt | |

## Live Foods

*Portion size: 1 cup*
*Frequency: At least 4–5 servings daily*

| Superfoods to Emphasize | Toxins to Limit or Avoid |
| --- | --- |
| Alfalfa sprouts | Aloe vera ● |
| Artichoke | Asparagus ● |
| Arugula | Beet greens |
| Avocado ◊ | Cassava |
| Beet ◊ | Cauliflower ● |
| Bok choy, pak choi ◊ | Celery |

| Superfoods to Emphasize | Toxins to Limit or Avoid |
| --- | --- |
| Brussels sprout ◊ | Chayote ● |
| Cabbage | Chicory |
| Carrot ◊ | Collard greens |
| Chinese kale, kai-lan ◊ | Corn, popcorn |
| Dandelion | Cucumber ● |
| Escarole ◊ | Daikon radish |
| Fiddlehead fern ◊ | Eggplant |
| Ginger | Endive ● |
| Grape leaves ◊ | Fenugreek ● |
| Kale ◊ | Jicama |
| Hearts of palm | Kohlrabi |
| Horseradish | Lettuce, green leaf, iceberg |
| Leek ◊ | Mushroom, oyster ● |
| Mushroom, enoki | Mushroom, portobello ● |
| Mushroom, commercial ◊ | Mushroom, shiitake |
| Mushroom, maitake | Mushroom, straw ● |
| Mustard greens | Okra ● |
| Onion, all types ◊ | Olive, black |
| Pumpkin ◊ | Olive, green ● |
| Radish sprouts | Parsnip ● |
| Rutabaga ◊ | Pepper, bell ● |
| Sauerkraut | Pepper, chili, jalapeño |
| Scallion | Pickle, vinegar |
| Sea vegetables, Irish moss ◊ | Potato, white with skin |
| Sea vegetables, kelp | Purslane |
| Sea vegetables, wakame | Quorn ● |
| Squash | Radish ● |
| Swiss chard ◊ | Rhubarb |
| Tomatillo | Sea vegetables, agar |
| Turnip greens ◊ | Spinach |
| Watercress ◊ | Sweet potato |
| Zucchini | Tomato ● |
| | Turnip ● |
| | Water chestnut ● |
| | Yam |

# Fruits

*Portion size: 1 cup fruit or 1 medium–sized fruit*
*Frequency: At least 3 servings daily*

| Superfoods to Emphasize | Toxins to Limit or Avoid |
|---|---|
| Blueberry | Apple ● |
| Cherimoya | Apricot ● |
| Cloudberry | Banana |
| Cranberry ◊ | Bitter melon |
| Currants ◊ | Blackberry ● |
| Date ◊ | Cherry ● |
| Dewberry | Coconut meat |
| Durian ◊ | Grape |
| Elderberry | Guava ● |
| Goji berry | Honeydew |
| Gooseberry | Kumquat |
| Grapefruit | Loquat |
| Groundcherry | Mango ● |
| Kiwi | Noni |
| Lemon ◊ | Orange |
| Lime ◊ | Peach ● |
| Lingonberry ◊ | Pear ● |
| Loganberry ◊ | Pomegranate |
| Mamey sapote | Raisin |
| Mulberry | Strawberry ● |
| Muskmelon | Tangerine |
| Nectarine ◊ | Watermelon ● |
| Papaya | |
| Passion fruit | |
| Pawpaw | |
| Persimmon | |
| Pineapple ◊ | |
| Quince | |
| Raspberry | |
| Rowanberry ◊ | |
| Tamarillo | |
| Youngberry | |

## Spices

*Portion size: 1 teaspoon*
*Frequency: At least 1–2 servings daily*

| Superfoods to Emphasize | Toxins to Limit or Avoid |
|---|---|
| Allspice | Acacia (gum arabic) |
| Arrowroot | Anise ● |
| Basil ◊ | Celery seed ● |
| Bay leaf | Chervil ● |
| Caper | Chili powder |
| Cilantro ◊ | Chives |
| Cinnamon ◊ | Chocolate |
| Curry ◊ | Cornstarch |
| Dill | Cream of tartar |
| Fennel | Cumin ● |
| Garlic ◊ | Licorice root |
| Nutmeg | Marjoram ● |
| Oregano ◊ | Pepper, black |
| Paprika | Savory ● |
| Peppermint | Tarragon ● |
| Rosemary ◊ | Wintergreen |
| Sage | |
| Thyme | |
| Turmeric ◊ | |

## Beverages

*Portion size: 6–8 ounces*
*Frequency 2–4 servings daily*

| Superfoods to Emphasize | Toxins to Limit or Avoid |
|---|---|
| Cherry juice | Apple juice ● |
| Coffee | Beer |
| Cranberry juice | Blackberry juice |
| Cucumber juice | Celery juice |

| Superfoods to Emphasize | Toxins to Limit or Avoid |
|---|---|
| Elderberry juice ◊ | Cola beverages |
| Grapefruit juice ◊ | Diet sodas |
| Lemon and water ◊ | Liquor, distilled |
| Milk, almond ◊ | Milk, coconut |
| Pineapple juice ◊ | Milk, rice ● |
| Tea, black | Milk, soy ● |
| Tea, chamomile | Orange juice |
| Tea, gingerroot | Pear juice |
| Tea, ginseng ◊ | Prune juice |
| Tea, green, kukicha ◊ | Seltzer water ● |
| Tea, roselle | Tangerine juice |
| | Tomato juice |
| | Watermelon juice ● |
| | Wine, white |

## Condiments and Additives

*Portion size: 1 teaspoon*
*Frequency: Use as needed*

| Superfoods to Emphasize | Toxins to Limit or Avoid |
|---|---|
| Barley malt | Aspartame |
| Epazote | Carrageenan |
| Fruit pectin ◊ | Dextrose |
| Honey ◊ | Fructose |
| Locust bean gum | Gelatin, plain |
| Mastic gum | Guar gum |
| Mayonnaise, tofu | High-fructose corn syrup |
| Mustard powder | Ketchup |
| Reihan, Tulsi (Holy basil) | Konjac |
| Sea salt | MSG |
| Soybean sauce, tamari | Maple syrup ● |
| Vegetable glycerine ◊ | Mayonnaise |
| Yeast extract spread, Marmite | Miso ● |
| Yeast, baker's ◊ | Molasses ● |
| Yeast, nutritional ◊ | Molasses, blackstrap ● |

| Superfoods to Emphasize | Toxins to Limit or Avoid |
|---|---|
| | Mono- and diglycerides |
| | Mustard, with vinegar |
| | Phosphoric acid |
| | Pickle relish |
| | Pimaricin (natamycin) |
| | Polyvinyl pyrrolidone |
| | Potassium bisulfite |
| | Potassium metabisulfite |
| | Propionic acid |
| | Rice syrup |
| | Sodium nitrite |
| | Sodium sulfite |
| | Sugar, brown, white |
| | Sulfur dioxide |
| | Vinegar, all types |
| | Worcestershire sauce |

## GenoType Teacher Supplement Guide

These supplements can help improve your results while on the Teacher Diet. Most are readily available in health food stores, while a few are a bit more obscure. Note that these supplements are solely recommended for use as part of the Teacher GenoType Diet program. To examine how other supplements you may be using rate with GenoType Teacher, or for more detailed information on the science behind the use of this supplement protocol, visit the official GenoType Diet Web site at www.genotypediet.com. Remember always to discuss the use of any nutritional supplements with your physician before embarking on a supplement program.

*Supplements that help Teachers control bacterial overgrowth and help regulate the metabolism:*
- Thiamine (vitamin B1). Typical daily dose: 25–50 mg
- Biotin (vitamin B7). Typical daily dose: 1–4 mg

- Bifidobacteria probiotics. Typical daily dose: 1.5 billion CFU
- Brewer's yeast (*Saccharomyces cerevisiae*). Typical daily dose: 1 tsp

*Supplements that enhance the Teacher's epigenetic functions (DNA methylation and histone acetylation):*
- Curcumin (turmeric extract). Typical daily dose: 200–500 mg
- Betaine hydrochloride. Typical daily dose: 1–2 grams with food
- Selenium. Typical daily dose: 25–100 mcg

*Supplements that stabilize the Teacher's genetic integrity and block gene mutations:*
- Quercetin. Typical daily dose: 200–1000 mg
- Vitamin D (cholecalciferol). Typical daily dose : 400–800 IU
- Green tea. Typical daily dose: 2–3 cups of tea
- Ginseng (*Panax* species), standardized to provide at least 10% ginsenosides. Typical daily dose: 100–300 mg

## GenoType Teacher Lifestyle
### Meal Planning

The Teacher has a range of food combinations and cuisines available for imaginative meal planning. We've collected a variety of meal plans (even plans for families with different member GenoTypes) and hundreds of tasty recipes for each GenoType online at www.genotypediet.com.

### Exercise Guide

Teacher GenoTypes need regular, vigorous exercise to keep their metabolism in balance, increase endurance, and maximize their resistance. You'll need to blend both *demanding* and *less demanding* exercises for a total of 40 minutes, 4–5 times weekly. Be sure to warm up for at least 5–10 minutes with some gentle stretching before beginning any aerobic activity.

*Less Demanding*
- Pilates or other forms of core strengthening
- Chi Gong or tai chi
- Light upper- and lower-body weight work
- Yoga

*More Demanding*
- Hiking
- Vigorous walking
- Moderate competitive sports (tennis, racquetball, volleyball)
- Moderate resistance training

*Purify and Tonify*

As the characteristics of this GenoType began to coalesce, I could identify an almost cultural element to the Teacher: a lifestyle and set of personal habits that appear spartan, to say the least. One could see the early teachers performing *Misogi,* a practice involving purification before sunrise with cold water, often under a waterfall. Before entering the waterfall, people would attempt to raise their metabolism and absorb as much Chi as possible by special types of deep breathing. Paradoxically, although it can be quite helpful for them, many Teachers, especially when stressed or high-strung, are not particularly drawn to cold baths, showers, or iced drinks. If you, like many people, don't have a waterfall handy and nearby, consider ending your morning shower with cold water. It may take some getting used to, but you may find yourself getting addicted to it!

### How to Do It
- Stand in the shower with your feet fairly close together. Start with a tepid shower water temperature. Relax and free your mind from all worry and distractions.
- Open your mouth slightly.
- Press gently and smoothly inward on your abdomen with your palms. Breathe out during a count to 6.
- Hold for a count of 2.

- Release smoothly and breathe in during a count to 6.
- Hold for a count of 2.
- Repeat this whole process three times.
- Drop your arms to your sides and close your eyes if they are not already closed.
- Let go of everything and enjoy the state of "nothingness" for about 5 minutes or the time that feels right for you.
- Lower the water temperature of the shower until it is just a bit uncomfortable.
- Repeat this whole process three times.
- End by giving yourself a brisk head-to-toe body rub with a coarse towel.

# CHAPTER EIGHTEEN

# The Explorer Diet

**W**elcome, Explorer! This chapter contains all the information that you'll need to get started on the diet for your GenoType. Diet buddy support, new research, and help with recipes and meal planning are only a mouse click away, at the official GenoType Diet Web site (www.genotypediet.com).

The Explorer Diet is broken down by food category. Each category (Red Meats, Poultry, etc.) contains two lists. The list on the left contains the *Explorer Superfoods*, which are foods that in the Explorer body act as medicines, balancing stress, regenerating genes, and repairing the digestive tract. Foods that help Explorers maintain their ideal weight, increase muscle mass, and decrease body fat are identified by a diamond (◊) icon. To gain the maximum benefit from the Explorer Diet, they should be routinely consumed.

The list on the right contains the *Explorer Toxins*, foods that GenoType Explorer would be wise to avoid. Some foods on the Explorer Toxin list need only be avoided for a short period of time so that you can regain your balance. After 3–6 months, you can reintroduce these foods back into your diet in modest amounts. They are identified by a black dot (●) icon.

Of course, if you are battling an illness or feel your weight beginning to creep back up, you may want to ramp up your compliance by avoiding these foods again for a while.

If a food is not listed, it is essentially *neutral*, meaning that the nutrients in it will benefit you but won't specifically help you restore balance to your genes or health to your cells. Feel free to eat these foods—but don't neglect the foods I recommend. A complete list of all the foods I've tested is available online (www.genotypediet.com).

## Red Meats

*Portion size: About the size of your hand (4–6 ounces)*
*Frequency: 2–3 times weekly*

| Superfoods to Emphasize | Toxins to Limit or Avoid |
|---|---|
| Calf liver | Beef heart |
| Goat | Beef tongue |
| Lamb ◊ | Boar |
| Mutton | Ham |
| Rabbit | Horse |
| Soup, bone | Pork, bacon |
| Soup, marrow ◊ | Sweetbreads |

## Poultry

*Portion size: About the size of your hand (4–6 ounces)*
*Frequency: 1–3 times weekly*

| Superfoods to Emphasize | Toxins to Limit or Avoid |
|---|---|
| Emu ◊ | Chicken liver |
| Ostrich ◊ | Duck |
| Partridge ◊ | Duck liver |
| Pheasant | Goose liver ● |
| Quail ◊ | |
| Squab | |
| Turkey | |

## Fish and Seafood

*Portion size: About the size of your hand (4–6 ounces)*
*Frequency: No more than 3 times weekly for first three months of program;*
*4 times weekly thereafter*

| Superfoods to Emphasize | Toxins to Limit or Avoid |
|---|---|
| Bullhead ◊ | Bass, blue gill |
| Butterfish | Bass, sea, lake |
| Carp | Bass, striped |
| Catfish ◊ | Bluefish |
| Chub | Clams |
| Cod | Crab |
| Croaker ◊ | Eel |
| Cusk | Flounder |
| Drum ◊ | Haddock ● |
| Gray sole | Halibut ● |
| Grouper | Lobster ● |
| Halfmoon fish | Mackerel, Atlantic |
| Harvest fish | Mackerel, Spanish |
| Monkfish | Mullet ● |
| Ocean pout | Mussels |
| Opaleye fish | Orange roughy |
| Pompano ◊ | Pike |
| Scrod | Salmon, chinook ● |
| Sea bream ◊ | Salmon, sockeye ● |
| Shad | Salmon, Atlantic, farmed |
| Sturgeon | Salmon, Atlantic, wild |
| Sucker ◊ | Shrimp |
| Snail, escargot ◊ | Red snapper |
| Turbot, European ◊ | Tilapia |
| Whitefish ◊ | Tilefish ● |
| Wolffish, Atlantic ◊ | Trout, rainbow, wild ● |
| | Trout, sea |
| | Trout, steelhead, wild ● |
| | Tuna, bluefin |
| | Tuna, skipjack |
| | Tuna, yellowfin |
| | Whiting ● |
| | Yellowtail |

## Eggs and Roes

*Portion size: 1 egg*
*Frequency: 5–7 servings per week*

| Superfoods to Emphasize | Toxins to Limit or Avoid |
| --- | --- |
| Goose egg ◊ | Carp roe |
| Quail egg | Egg white, chicken ● |
| Sailfish roe | Egg white, duck |
| | Egg, whole, chicken ● |
| | Egg, whole, duck |
| | Egg yolk, chicken ● |

## Dairy

*Portion size: Milk: 6 ounces; cheese: 2–4 ounces*
*Frequency: Cheeses: 3–5 times weekly; ghee as desired*

| Superfoods to Emphasize | Toxins to Limit or Avoid |
| --- | --- |
| Ghee (clarified butter) ◊ | American cheese |
| Mozzarella cheese | Blue cheese |
| Paneer cheese | Brie cheese |
| Ricotta cheese | Buttermilk, low-fat ● |
| Romanian Urdă | Camembert cheese |
| Whey protein powder | Casein |
| | Cheddar cheese |
| | Cheshire cheese |
| | Colby cheese |
| | Cottage cheese ● |
| | Edam cheese |
| | Farmer's cheese ● |
| | Feta cheese ● |
| | Gorgonzola cheese |
| | Gouda cheese |
| | Gruyère cheese |
| | Half-and-half |
| | Havarti cheese |

| Superfoods to Emphasize | Toxins to Limit or Avoid |
|---|---|
| | Jarlsberg cheese |
| | Kefalotyri cheese ● |
| | Kefir ● |
| | Limburger cheese |
| | Manchego cheese |
| | Milk, cow, skim or 2% |
| | Milk, cow, whole |
| | Milk, goat |
| | Monterey Jack cheese |
| | Muenster cheese |
| | Neufchâtel cheese |
| | Parmesan cheese |
| | Pecorino cheese |
| | Port du Salut cheese |
| | Provolone cheese |
| | Quark cheese ● |
| | Romano cheese |
| | Roquefort cheese |
| | Stilton cheese |
| | Yogurt ● |

## Vegetable Proteins

*Portion size: Nuts, seeds: ½ cup; nut butters: 2 tablespoons*
*Frequency: 3–7 times weekly*

| Superfoods to Emphasize | Toxins to Limit or Avoid |
|---|---|
| Adzuki bean | Almond ● |
| Butter bean ◊ | Almond butter ● |
| Cannellini bean | Beechnut |
| Carob ◊ | Black bean ● |
| Chestnut, Chinese | Black-eyed pea ● |
| Chia seed, pinole ◊ | Brazil nut ● |
| Great northern bean ◊ | Broad bean, fava |
| Green, string bean ◊ | Butternut |
| Haricot bean | Cashew |

| Superfoods to Emphasize | Toxins to Limit or Avoid |
|---|---|
| Hickory nuts | Cashew butter |
| Lentils, all types ◊ | Chestnut, European ● |
| Lentils, sprouted ◊ | Copper bean ● |
| Litchi | Fava bean |
| Lotus root | Filbert, hazelnut |
| Lotus seeds | Flaxseed, linseed ● |
| Macadamia ◊ | Garbanzo bean, chickpea |
| Moth bean | Hemp seed |
| Peas ◊ | Kidney bean |
| Pecan | Lima bean |
| Pine nut, pignolia | Lima bean flour |
| Pinto bean ◊ | Lupin seeds |
| Pinto beans, sprouted | Mung bean |
| Sapodilla | Natto |
| Snap bean ◊ | Navy bean ● |
| Winged beans | Peanut butter |
| Yard-long bean | Peanut flour |
| | Peanuts |
| | Pistachio |
| | Poppy seed ● |
| | Pumpkin seed |
| | Safflower seed |
| | Sesame butter, tahini ● |
| | Sesame seed ● |
| | Sesame flour |
| | Soybean |
| | Soybean, miso ● |
| | Soybean pasta |
| | Soybean, sprouted |
| | Soybean, tempeh |
| | Soybean, tofu |
| | Sunflower seed |
| | Tamarind bean |
| | Tempeh |
| | Walnut ● |
| | Watermelon seeds |
| | White bean |
| | Yeast, nutritional ● |

## Fats and Oils

*Portion size: 1 tablespoon*
*Frequency: 3–9 times weekly*

| Superfoods to Emphasize | Toxins to Limit or Avoid |
|---|---|
| Babassu oil | Almond oil ● |
| Camelina oil ◊ | Avocado oil |
| Chia seed oil | Borage seed oil |
| Ghee (clarified butter) | Butter |
| Grape seed oil | Canola oil |
| Hemp seed oil | Coconut oil |
| Herring oil | Cod liver oil |
| Macadamia oil ◊ | Corn oil |
| Oat oil | Cottonseed oil |
| Olive oil | Flaxseed, linseed oil ● |
| Perilla seed oil ◊ | Hazelnut nut oil |
| Quinoa oil ◊ | Lard |
| Rice bran oil ◊ | Margarine |
| Salmon oil | Palm oil |
| Shea nut oil ◊ | Peanut oil |
| Tea seed oil | Safflower oil |
| | Sesame oil ● |
| | Soybean oil |
| | Sunflower oil |
| | Walnut oil |
| | Wheat germ oil |

## Carbohydrates

*Portion size: ½ cup grains, cereals, rice; ½ muffin; 1 slice bread*
*Frequency: 2–3 times daily*

| Superfoods to Emphasize | Toxins to Limit or Avoid |
|---|---|
| 100% artichoke flour/pasta | Barley |
| Amaranth | Buckwheat, kasha, soba ● |
| Essene, Manna bread | Cornmeal, hominy, polenta ● |

| Superfoods to Emphasize | Toxins to Limit or Avoid |
| --- | --- |
| Fonio ◊ | Flaxseed bread ● |
| Job's tears ◊ | Oat bran ● |
| Kudzu | Oat flour ● |
| Lentil flour, dhal, poppadom | Rye ● |
| Millet ◊ | Sorghum |
| Quinoa ◊ | Soybean flour |
| Rice bran ◊ | Taro |
| Rice flour, brown ◊ | Wheat, bran, germ |
| Rice flour, white | Wheat, bulgur |
| Rice, basmati ◊ | Wheat, durum |
| Rice, brown ◊ | Wheat, emmer, einkorn |
| Rice, white | Wheat, kamut ● |
| Rice, wild | Wheat, white flour |
| Tapioca, manioc, cassava | Wheat, whole-grain flour |
| Teff | |
| Wheat, 100% sprouted | |
| Wheat, spelt | |

## Live Foods

*Portion size: 1 cup*
*Frequency: At least 4–5 servings daily*

| Superfoods to Emphasize | Toxins to Limit or Avoid |
| --- | --- |
| Artichoke ◊ | Alfalfa sprouts |
| Arugula | Aloe vera |
| Asparagus | Avocado ● |
| Asparagus pea | Brussels sprout |
| Bamboo shoot | Cabbage |
| Beet | Cauliflower |
| Beet greens | Corn, popcorn ● |
| Bok choy, pak choi | Eggplant |
| Borage | Fenugreek ● |
| Broccoflower | Leek ● |
| Broccoli | Mushroom, commercial |
| Broccoli leaves | Mushroom, oyster |

| Superfoods to Emphasize | Toxins to Limit or Avoid |
|---|---|
| Broccoli rabe, rapini | Mushroom, portobello |
| Carrot | Mushroom, shiitake |
| Chicory | Mushroom, straw |
| Chicory root ◊ | Olive, black |
| Chinese kale, kai-lan ◊ | Olive, green |
| Collard greens ◊ | Pepper, bell ● |
| Daikon radish ◊ | Pepper, chili, jalapeño ● |
| Dandelion ◊ | Pickle, brine |
| Escarole ◊ | Pickle, vinegar |
| Fiddlehead fern | Potato, white with skin |
| Garlic ◊ | Quorn |
| Ginger ◊ | Rhubarb ● |
| Grape leaves ◊ | Sauerkraut |
| Hearts of palm | Sea vegetables, agar |
| Horseradish ◊ | Spinach ● |
| Jerusalem artichoke ◊ | Sweet potato ● |
| Jute, potherb | Sweet potato leaves ● |
| Kale | Tomatillo ● |
| Kanpyo ◊ | Tomato ● |
| Kohlrabi ◊ | Water chestnut |
| Lettuce, green leaf, iceberg | Yam ● |
| Lettuce, romaine | |
| Mustard greens | |
| Okra | |
| Onion, all types ◊ | |
| Oyster plant, salsify | |
| Parsnip ◊ | |
| Radish | |
| Radish leaf ◊ | |
| Radish sprouts ◊ | |
| Rutabaga ◊ | |
| Scallion | |
| Sea vegetables, kelp | |
| Sea vegetables, spirulina | |
| Sea vegetables, wakame | |
| Shallots | |
| Swamp cabbage, skunk cabbage | |
| Swiss chard ◊ | |

| Superfoods to Emphasize | Toxins to Limit or Avoid |
| --- | --- |
| Taro, Tahitian<br>Turnip<br>Turnip greens ◊<br>Watercress | |

## Fruits

*Portion size: 1 cup fruit or 1 medium-sized fruit*
*Frequency: At least 3 servings daily*

| Superfoods to Emphasize | Toxins to Limit or Avoid |
| --- | --- |
| Acai berry ◊ | Apricot ● |
| Apple | Asian pear |
| Breadfruit ◊ | Banana ● |
| Canistel | Bitter melon |
| Carissa, natal plum | Blackberry |
| Cherimoya | Blueberry ● |
| Cloudberry | Cantaloupe melon |
| Cranberry ◊ | Casaba melon |
| Currants ◊ | Coconut meat |
| Elderberry ◊ | Cherry ● |
| Feijoa | Date |
| Goji, wolfberry | Durian ● |
| Gooseberry ◊ | Fig ● |
| Grapefruit | Grape ● |
| Groundcherry | Honeydew |
| Guava ◊ | Kiwi |
| Java-plum | Loquat |
| Jujube | Mango ● |
| Kumquat | Muskmelon ● |
| Lemon | Nectarine ● |
| Lime | Orange ● |
| Lingonberry ◊ | Papaya ● |
| Loganberry | Peach |
| Mamey sapote | Pear ● |
| Mulberry | Plantain |

| Superfoods to Emphasize | Toxins to Limit or Avoid |
| --- | --- |
| Passion fruit | Plum |
| Pawpaw ◊ | Prune |
| Persian melon ◊ | Raisin |
| Persimmon | Strawberry |
| Pineapple | Tangerine ● |
| Prickly pear | |
| Pomegranate | |
| Pummelo ◊ | |
| Quince ◊ | |
| Raspberry ◊ | |
| Rowanberry ◊ | |
| Sapotes | |
| Spanish melon | |
| Star fruit, carambola | |
| Watermelon | |

# Spices

*Portion size: 1 teaspoon*
*Frequency: At least 1–2 servings daily*

| Superfoods to Emphasize | Toxins to Limit or Avoid |
| --- | --- |
| Cilantro ◊ | Acacia (gum arabic) |
| Curry | Allspice |
| Fennel | Anise |
| Garlic ◊ | Caper |
| Thyme ◊ | Caraway |
| Turmeric | Chocolate ● |
| | Cinnamon |
| | Guarana |
| | Mace |
| | Pepper, black |
| | Senna |
| | Vanilla |

# Beverages

*Portion size: 6–8-ounces*
*Frequency: 2–4 servings daily*

| Superfoods to Emphasize | Toxins to Limit or Avoid |
|---|---|
| Juice, apple ◊ | Beer |
| Juice, cranberry | Coffee |
| Juice, goji berry | Coffee, decaffeinated ● |
| Juice, lemon and water | Cola beverages |
| Juice, pomegranate | Diet sodas |
| Tea, gingerroot | Juice, blackberry |
| Tea, green, kukicha | Juice, grape ● |
| Tea, licorice root ◊ | Juice, orange ● |
| Tea, peppermint | Juice, pear |
| Tea, rose hip ◊ | Juice, prune |
| Tea, verbena ◊ | Juice, tangerine ● |
|  | Juice, tomato |
|  | Liquor, distilled |
|  | Milk, coconut |
|  | Milk, rice |
|  | Milk, soy |
|  | Tea, black |
|  | Tea, black, decaffeinated ● |
|  | Wine, red ● |
|  | Wine, white ● |

# Condiments and Additives

*Portion size: 1 teaspoon*
*Frequency: Use as needed*

| Superfoods to Emphasize | Toxins to Limit or Avoid |
|---|---|
| Agave syrup | Acetaminophen |
| Epazote | Aspartame |
| Fruit pectin | Barley malt ● |
| Honey | Caffeine |

| Superfoods to Emphasize | Toxins to Limit or Avoid |
|---|---|
| Konjac | Calcium diglutamate |
| Locust bean gum | Calcium EDTA |
| Maple syrup | Carrageenan |
| Molasses ◊ | Clonazepam |
| Molasses, blackstrap ◊ | Disodium ribonucleotide |
| Reihan, Tulsi (Holy basil) | Food dyes and colorings |
| Rice syrup | Fructose |
| Roselle | Gelatin, plain |
| Sea salt | Glutamic acid |
| Vegetable glycerine | Guar gum |
| | High-fructose corn syrup |
| | Ketchup |
| | Mayonnaise |
| | Mayonnaise, tofu |
| | Monoammonium glutamate |
| | Monopotassium glutamate |
| | MSG (monosodium glutamate) |
| | Mustard, with vinegar ● |
| | Mustard powder ● |
| | Pickle relish |
| | Polysorbate |
| | Potassium bisulfite |
| | Potassium metabisulfite |
| | Sodium sulfite |
| | Soybean sauce, tamari |
| | Stevia ● |
| | Sugar, brown, white |
| | Sulfasalazine |
| | Sulfur dioxide |
| | Umeboshi plum, vinegar ● |
| | Vinegar, red wine ● |
| | Vinegar, white |
| | Worcestershire sauce |
| | Yeast extract spread, Marmite ● |
| | Yeast, nutritional ● |

# GenoType Explorer Supplement Guide

These supplements can help improve your results while on the Explorer Diet. Most are readily available in health food stores, while a few are a bit more obscure. Note that these supplements are solely recommended for use as part of the Explorer GenoType Diet program. To examine how other supplements you may be using rate with GenoType Explorer, or for more detailed information on the science behind the use of this supplement protocol, visit the official GenoType Diet Web site at www.genotypediet.com. Remember to discuss the use of any nutritional supplements with your physician before embarking on a supplement program.

*Supplements that help Explorers tune up the "detoxification genes":*
- Milk thistle (silymarin). Typical daily dose: 200–500 mg
- N-acetyl glucosamine (NAG). Typical daily dose: 200–500 mg
- Artichoke leaf extract (10–20% chlorogenic acid). Typical daily dose: 100–250 mg

*Supplements that tone down the Explorer's "chemical sensitivity genes" and help regulate the metabolism:*
- Calcium d-glucarate. Typical daily dose: 200–500 mg
- Plant sterols (sterolins). Typical daily dose: 25–100 mg
- Reduced glutathione. Typical daily dose: 100–500 mg

*Supplements that strengthen the Explorer's blood, bone marrow, and liver:*
- Licorice extract. Typical daily dose: 100–200 mg*
- Curcumin extract. Typical daily dose: 200–500 mg
- Drynaria root. Typical daily dose: 100–500 mg
- Larch arabinogalactan powder. Typical daily dose: 100–300 mg

---

* If you have high blood pressure or retain water, you should use whole-licorice preparations only under a physician's supervision. You can, however, use the DGL (deglycerinated) form of licorice (available at many health food stores and pharmacies) if you wish.

# GenoType Explorer Lifestyle

## Meal Planning

The Explorer has a range of food combinations and cuisines available for imaginative meal planning. We've collected a variety of meal plans (even plans for families with different member GenoTypes) and hundreds of tasty recipes for each GenoType online at www.genotypediet.com.

## Exercise Guide

It should be emphasized that the one condition of all exercise for Explorers is the breaking of a sweat. If you are an Explorer and you do not sweat when you exercise, you are not exercising enough! To stay fit and healthy, reduce stress, and increase endurance, you'll need about 40 minutes of exercise 4–5 times daily. Be sure to warm up for at least 5–10 minutes with some gentle stretching before beginning any aerobic activity.

*Less Demanding*
- Hiking
- Pilates or other forms of core strengthening
- Vigorous walking
- Moderate competitive sports (tennis, racquetball, volleyball)
- Light upper- and lower-body weight work

*More Demanding*
- Aerobics
- Intense competitive sports (martial arts, basketball, soccer)
- Dancing
- Gymnastics
- Moderate resistance training
- Running

*Give yourself an "overnight detox" every month:*
Explorers benefit from regular, gentle detoxification cleanses. A simple monthly tune-up for the liver and gallbladder involves a time-honored technique called the castor oil pack, followed by a good sweat.

## How to Do It

- The day before you do your Overnight Detox, make a point of drinking an 8-ounce glass of organic apple juice every 3 hours for a maximum of 5 glasses.
- Eat only light foods after midday, such as salads, juices, and fruits.
- One hour before bed, take 2 tablespoons of olive oil, followed by 2 tablespoons of lemon or lime juice.*
- Now make your castor oil pack. The pack is made by soaking a piece of flannel in castor oil, then putting it in the area of the liver (the right side of your abdomen, under the bottom rib) and placing a heat source, such as a hot water bottle, on top of it. You can set up the castor oil pack to work overnight by using an electric blanket–type heater set to low.
- Next morning, you might be surprised to see green blobs in your bowel movement. Many people think these are gallstones, but they are not. However, they do afford the liver a means of removing fat-soluble toxins.
- If you have access to a steam room and can get a 20-minute steam in, go for it.

---

* If you have gallstones or suffer from irritable-bowel problems, you can skip this step.

# CHAPTER NINETEEN

# The Warrior Diet

**W**elcome, Warrior! This chapter contains all the information you'll need to get started on the diet for your GenoType. Diet buddy support, new research, and help with recipes and meal planning are only a mouse click away, at the official GenoType Diet Web site (www.genotypediet.com).

The Warrior Diet is broken down by food category. Each category (Red Meats, Poultry, etc.) contains two lists. The list on the left contains the *Warrior Superfoods,* foods that act as medicines, balancing stress, regenerating genes, and repairing the digestive tract. Superfoods that help Warriors maintain ideal weight, increase muscle mass, and decrease body fat are identified by a diamond (◊) icon. To gain the maximum benefit from the Warrior Diet, these foods should be routinely consumed.

The list on the right contains the *Warrior Toxins,* foods that Warrior GenoTypes would be wise to avoid. Some foods on the Warrior Toxin list need only be avoided for a short period of time so that you can regain your balance. After 3–6 months, you can reintroduce these foods back into your diet in modest amounts. They are identified by a black dot (•) icon. Of

course, if you are battling an illness or feel your weight beginning to creep back up, you may want to ramp up your compliance by avoiding these foods again for a while.

If a food is not listed, it is essentially *neutral*, meaning that the nutrients in it will benefit you but won't specifically help you restore balance to your genes or health to your cells. Feel free to eat these foods—but don't neglect the foods I recommend. A complete list of all the foods I've tested is available online (www.genotypediet.com).

## Red Meats

*Portion size: About the size of your hand (4–6 ounces)*
*Frequency: 0–1 times weekly*

| Superfoods to Emphasize | Toxins to Limit or Avoid |
|---|---|
| *There are no recommended red meats.* | Beef, bone soups, and broths |
| | Beef heart |
| | Beef liver |
| | Boar |
| | Buffalo, bison |
| | Goat |
| | Ham |
| | Lamb |
| | Mutton |
| | Pork |
| | Pork, bacon |
| | Sweetbreads |
| | Venison |

# Poultry

*Portion size: About the size of your hand (4–6 ounces)*
*Frequency: 0–2 times weekly*

| Superfoods to Emphasize | Toxins to Limit or Avoid |
|---|---|
| *There are no recommended poultry items.* | Chicken |
| | Chicken liver |
| | Cornish hen |
| | Duck |
| | Duck liver |
| | Goose |
| | Goose liver |
| | Grouse |
| | Guinea hen |
| | Partridge |
| | Pheasant |
| | Quail |
| | Squab |

# Fish and Seafood

*Portion size: About the size of your hand (4–6 ounces)*
*Frequency: 3–4 times weekly*

| Superfoods to Emphasize | Toxins to Limit or Avoid |
|---|---|
| Anchovy ◊ | Barracuda |
| Bullhead | Bass, sea, lake |
| Cod ◊ | Bass, striped |
| Grouper ◊ | Bluefish |
| Haddock | Catfish |
| Hake | Chub |
| Halfmoon fish | Clam |
| Halibut | Conch |
| Harvest fish | Crab |
| Mahimahi ◊ | Eel |
| Monkfish | Flounder |

| Superfoods to Emphasize | Toxins to Limit or Avoid |
| --- | --- |
| Mullet | Herring |
| Muskellunge | Lobster |
| Mussels | Mackerel, Atlantic |
| Ocean pout | Mackerel, Spanish |
| Octopus | Orange roughy |
| Opaleye fish | Oyster |
| Perch | Pompano |
| Pickerel (walleye) ◊ | Salmon, Atlantic, farmed |
| Pike | Salmon, Atlantic, wild |
| Pilchard | Sardine ● |
| Porgy | Shark |
| Red snapper ◊ | Shrimp |
| Salmon, Alaskan ◊ | Skate |
| Salmon, sockeye ◊ | Smelt ● |
| Scrod ◊ | Sole |
| Scup | Squid, calamari |
| Shad | Swordfish |
| Sheepshead fish | Trout, farmed |
| Snail, escargot ◊ | Trout, sea |
| Sturgeon | Weakfish |
| Sucker | |
| Tilapia | |
| Tilefish ◊ | |
| Trout, rainbow, wild | |
| Trout, steelhead, wild | |
| Tuna, skipjack | |
| Tuna, yellowfin | |
| Turbot, European | |
| Whitefish | |
| Whiting ◊ | |
| Wolffish, Atlantic | |
| Yellowtail ◊ | |

## Eggs and Roes

*Portion size: 1 egg*
*Frequency: 5–7 servings per week*

| Superfoods to Emphasize | Toxins to Limit or Avoid |
|---|---|
| Egg white, chicken ◊ | Carp roe |
| Egg, whole, chicken | Caviar |
| | Egg, whole, duck |
| | Goose egg |
| | Quail egg |
| | Sailfish roe |

## Dairy

*Portion size: Milk: 6 ounces; cheese: 2–4 ounces; ghee, butter, 1 tsp*
*Frequency: Cheeses: 4 times weekly*

| Superfoods to Emphasize | Toxins to Limit or Avoid |
|---|---|
| Cottage cheese | American cheese |
| Farmer cheese | Blue cheese |
| Kefir | Brie cheese |
| Paneer cheese | Camembert cheese |
| Quark cheese | Casein |
| Whey protein powder yogurt | Cheddar cheese |
| | Cheshire cheese |
| | Colby cheese |
| | Cream cheese |
| | Edam cheese |
| | Emmenthal, "Swiss" cheese |
| | Feta cheese ● |
| | Gorgonzola cheese |
| | Gouda cheese |
| | Gruyère cheese |
| | Half-and-half |
| | Havarti cheese |
| | Jarlsberg cheese |

| Superfoods to Emphasize | Toxins to Limit or Avoid |
| --- | --- |
| | Kefalotyri cheese |
| | Limburger cheese |
| | Manchego cheese |
| | Milk, buttermilk, low-fat ● |
| | Milk, cow, skim or 2% |
| | Milk, cow, whole |
| | Milk, goat |
| | Milk, Indian buffalo |
| | Monterey Jack cheese |
| | Mozzarella cheese |
| | Muenster cheese |
| | Neufchâtel cheese |
| | Parmesan cheese |
| | Pecorino cheese |
| | Port du Salut cheese |
| | Provolone cheese |
| | Ricotta cheese ● |
| | Romanian Urdă ● |
| | Romano cheese |
| | Roquefort cheese |
| | Sour cream |
| | Stilton cheese |
| | String cheese |

## Vegetable Proteins

*Portion size: Nuts, seeds: ½ cup; nut butters: 2 tablespoons*
*Frequency: At least twice daily*

| Superfoods to Emphasize | Toxins to Limit or Avoid |
| --- | --- |
| Adzuki bean ◊ | Brazil nut |
| Almond ◊ | Cashew ● |
| Almond butter ◊ | Cashew butter ● |
| Beechnut | Filbert, hazelnut ● |
| Broad bean, fava ◊ | Garbanzo bean, chickpea |
| Butter bean | Kidney bean ● |

| Superfoods to Emphasize | Toxins to Limit or Avoid |
|---|---|
| Cannellini bean | Lima bean |
| Flaxseed, linseed ◊ | Lima bean flour |
| Green, string bean ◊ | Macadamia ● |
| Haricot beans | Pistachio |
| Hemp seed ◊ | Safflower seed |
| Lentils, all types | Sesame butter, tahini ● |
| Lentils, sprouted | Sesame flour ● |
| Lotus seeds | Sesame seed ● |
| Lupin seeds | Sunflower seed |
| Natto ◊ | Tamarind bean |
| Navy bean | |
| Peanut butter ◊ | |
| Peanut flour ◊ | |
| Peanuts ◊ | |
| Peas ◊ | |
| Pecan ◊ | |
| Pine nut, pignolia ◊ | |
| Pinto bean | |
| Pinto beans, sprouted | |
| Poppy seed | |
| Pumpkin seed | |
| Soybean | |
| Soybean, edamame ◊ | |
| Soybean, miso ◊ | |
| Soybean, sprouted ◊ | |
| Soybean, tempeh ◊ | |
| Soybean, tofu | |
| Walnut ◊ | |
| Watermelon seeds | |
| White bean ◊ | |

## Fats and Oils

*Portion size: 1 tablespoon*
*Frequency: 3–6 times weekly*

| Superfoods to Emphasize | Toxins to Limit or Avoid |
|---|---|
| Almond oil ◊ | Avocado oil |
| Apricot kernel oil | Babassu oil |
| Black currant seed oil | Butter |
| Borage seed oil | Coconut oil |
| Camelina oil ◊ | Corn oil |
| Chia seed oil | Cottonseed oil |
| Cod liver oil | Grape seed oil |
| Evening primrose oil ◊ | Hazelnut nut oil |
| Flaxseed, linseed oil ◊ | Herring oil |
| Ghee (clarified butter) | Lard |
| Hemp seed oil ◊ | Macadamia oil ● |
| Olive oil | Margarine |
| Perilla seed oil ◊ | Oat oil ● |
| Pumpkin seed oil | Palm oil |
| Quinoa oil ◊ | Peanut oil ● |
| Salmon oil | Rice bran oil ● |
| Walnut oil ◊ | Safflower oil |
| Wheat germ oil ◊ | Sesame oil ● |
|  | Shea nut oil |
|  | Soybean oil ● |
|  | Sunflower oil |
|  | Tea seed oil |

## Carbohydrates

*Portion size: ½ cup grains, cereals, rice; ½ muffin; 1 slice bread*
*Frequency: 2–3 times daily*

| Superfoods to Emphasize | Toxins to Limit or Avoid |
|---|---|
| Amaranth | 100% artichoke flour, pasta ● |
| Barley ◊ | 100% sprouted bread ● |
| Flaxseed bread ◊ | Buckwheat, kasha, soba ● |

| Superfoods to Emphasize | Toxins to Limit or Avoid |
|---|---|
| Kudzu ◊ | Cornmeal, hominy, polenta |
| Lentil flour, dhal | Essene, Manna bread ● |
| Oat bran | Fonio |
| Oat flour | Job's tears |
| Poi | Millet |
| Poppadom | Rice bran ● |
| Quinoa | Rice flour, white |
| Rice, basmati | Rice, white |
| Rice, brown ◊ | Rice, wild |
| Rice flour, brown ◊ | Sorghum |
| Rye ◊ | Tapioca, manioc, cassava |
| Rye flour ◊ | Teff ● |
| Soybean flour | Wheat, durum, semolina |
| Wheat, bran | Wheat, white flour |
| Wheat, emmer, einkorn ◊ | |
| Wheat, kamut | |
| Wheat, spelt | |
| Wheat, sprouted | |
| Wheat, whole-grain | |

## Live Foods

*Portion size: 1 cup*
*Frequency: At least 4–5 servings daily*

| Superfoods to Emphasize | Toxins to Limit or Avoid |
|---|---|
| Alfalfa sprouts | Aloe vera |
| Artichoke ◊ | Asparagus |
| Asparagus pea | Avocado ● |
| Beet greens ◊ | Broccoli ● |
| Bok choy, pak choi | Cabbage ● |
| Borage | Chayote, pipinella |
| Broccoflower ◊ | Corn, popcorn |
| Broccoli leaves | Eggplant ● |
| Broccoli, Chinese | Fenugreek |
| Brussels sprout ◊ | Jerusalem artichoke |

| Superfoods to Emphasize | Toxins to Limit or Avoid |
| --- | --- |
| Cauliflower ◊ | Mushroom, shiitake |
| Celery | Mushroom, white, silver dollar |
| Celtuce | Olive, black |
| Chicory | Oyster plant, salsify |
| Chinese kale, kai-lan ◊ | Parsnip ● |
| Collard greens | Pepper, bell ● |
| Cucumber | Pepper, chili, jalapeño |
| Dandelion | Pickle, brine |
| Endive ◊ | Pickle, vinegar |
| Escarole ◊ | Potato, white with skin |
| Fiddlehead fern | Radish ● |
| Ginger | Radish sprouts ● |
| Grape leaves ◊ | Rhubarb |
| Jute, potherb | Sauerkraut |
| Kale ◊ | Sea vegetables, agar |
| Kanpyo (dried gourd strips) | Squash ● |
| Leek | Sweet potato |
| Lettuce, green leaf, iceberg | Tomatillo ● |
| Lettuce, romaine | Tomato ● |
| Mushroom, maitake | Yam |
| Mustard greens | |
| Okra ◊ | |
| Olive, green ◊ | |
| Onion, all types ◊ | |
| Pumpkin | |
| Purslane ◊ | |
| Quorn | |
| Rutabaga ◊ | |
| Scallion | |
| Sea vegetables, spirulina | |
| Spinach | |
| Swamp cabbage, skunk cabbage | |
| Sweet potato leaves | |
| Swiss chard | |
| Turnip greens | |
| Watercress | |

# Fruits

*Portion size: 1 cup fruit or 1 medium-sized fruit*
*Frequency: At least 3 servings daily*

| Superfoods to Emphasize | Toxins to Limit or Avoid |
|---|---|
| Apple | Banana |
| Apricot ◊ | Bitter melon |
| Blackberry | Blueberry ● |
| Breadfruit | Canistel |
| Cantaloupe melon | Cherry ● |
| Cranberry ◊ | Coconut meat |
| Currants | Date ● |
| Durian ◊ | Dewberry |
| Elderberry ◊ | Guava |
| Fig | Honeydew |
| Goji, wolfberry | Huckleberry |
| Gooseberry | Kiwi ● |
| Grape ◊ | Loganberry |
| Grapefruit ◊ | Loquat |
| Kumquat | Mango |
| Lemon | Muskmelon ● |
| Lime ◊ | Nectarine ● |
| Lingonberry ◊ | Orange |
| Noni | Pear ● |
| Pawpaw | Persian melon |
| Peach ◊ | Persimmon |
| Plum | Pineapple ● |
| Pummelo | Plantain |
| Strawberry | Pomegranate |
| Tamarillo | Prickly pear |
| | Prune ● |
| | Quince |
| | Raisin ● |
| | Sago palm |
| | Spanish melon ● |
| | Star fruit ● |
| | Tangerine |

## Spices

*Portion size: 1 teaspoon*
*Frequency: At least 1–2 servings daily*

| Superfoods to Emphasize | Toxins to Limit or Avoid |
| --- | --- |
| Bay leaf | Allspice |
| Chocolate ◊ | Anise ● |
| Cilantro | Caper |
| Cinnamon ◊ | Chili powder |
| Curry | Clove |
| Garlic ◊ | Guarana |
| Oregano ◊ | Mace ● |
| Parsley ◊ | Nutmeg ● |
| Rosemary | Paprika |
| Turmeric | Pepper, black |
|  | Pepper, red flakes |
|  | Sage ● |
|  | Wintergreen |

## Beverages

*Portion size: 6–8-ounces*
*Frequency: 2–4 servings daily*

| Superfoods to Emphasize | Toxins to Limit or Avoid |
| --- | --- |
| Blackberry juice ◊ | Beer |
| Celery juice | Blueberry juice |
| Coffee | Cabbage juice |
| Cranberry juice ◊ | Cherry juice ● |
| Elderberry juice ◊ | Cola beverages |
| Goji berry juice | Diet sodas |
| Grape juice ◊ | Liquor, distilled |
| Grapefruit juice | Milk, coconut |
| Lemon and water | Orange juice |
| Milk, almond | Pear juice ● |
| Milk, rice | Pomegranate juice |

| Superfoods to Emphasize | Toxins to Limit or Avoid |
|---|---|
| Pineapple juice | Seltzer water |
| Pummelo juice | Tangerine juice |
| Tea, black ◊ | Tomato juice ● |
| Tea, gingerroot | Watermelon juice |
| Tea, green, kukicha, bancha ◊ | Wine, white |
| Tea, yerba maté ◊ | |
| Wine, red | |

## Condiments and Additives

*Portion size: 1 teaspoon to 1 tablespoon*
*Frequency: Use as needed*

| Superfoods to Emphasize | Toxins to Limit or Avoid |
|---|---|
| Barley malt ◊ | Acacia (gum arabic) |
| Fruit pectin | Aspartame |
| Maple syrup | BHA, BHT |
| Mastic gum | Calcium EDTA |
| Mayonnaise, tofu | Carrageenan |
| Mustard, dry powder ◊ | Gelatin, plain |
| Sea salt | Guar gum |
| Soybean sauce, tamari | High-fructose corn syrup |
| Vegetable glycerine | Honey ● |
| Yeast, nutritional | Ketchup |
| | Mayonnaise |
| | Molasses ● |
| | Molasses, blackstrap ● |
| | Mono- and diglycerides |
| | Mustard, with vinegar and wheat |
| | Pickle relish |
| | Sodium nitrite |
| | Sodium sulfite |
| | Vinegar, all types |
| | Worcestershire sauce |

# GenoType Warrior Supplement Guide

These supplements can help improve your results while on the Warrior Diet. Most are readily available in health food stores, while a few are a bit more obscure. Note that these supplements are solely recommended for use as part of the Warrior GenoType Diet program. To examine how other supplements you may be using rate with GenoType Warrior, or for more detailed information on the science behind the use of this supplement protocol, visit the official GenoType Diet Web site at www.genotypediet.com. Remember to discuss the use of any nutritional supplements with your physician before embarking on a supplement program.

*Supplements that tone down the Warrior's rapid aging due to "frittering chromosomes":*
• Resveratrol. Typical daily dose: 250–500 mg
• Sprouted greens supplement. Typical daily dose: 200–500 mg
• Danshen (*Salvia miltiorrhiza*). Typical daily dose: 50–250 mg

*Supplements that calm the Warrior cardiovascular system, reducing inflammation and protecting the heart and arteries:*
• Gynostemma. Typical daily dose: 100–400 mg
• Selenium. Typical daily dose: 50–100 mcg
• Feverfew (parthenolide). Typical daily dose: 100–400 mg

*Supplements that help turn down those thrifty Warrior genetic tendencies and help regulate the metabolism and appetite-control center:*
• Lipoic acid. Typical daily dose: 50–100 mg
• Vitamin $B_6$ (pyridoxal phosphate). Typical daily dose: 50–100 mg
• Punarnava or santhi (*Boerhaavia diffusa*). Typical daily dose: 50–250 mg

# GenoType Warrior Lifestyle

## Meal Planning

The Warrior has a range of food combinations and cuisines available for imaginative meal planning. We've collected a variety of meal plans (even plans for families with different member GenoTypes) and hundreds of tasty recipes for each GenoType online at www.genotypediet.com.

## Exercise Guide

The Warrior exercise protocol is designed to increase muscle growth, develop and strengthen the cardiovascular system, and offset the Warrior's tendency to rapid aging. Warrior GenoTypes do best with stretching and lengthening exercises that also have a resistance component to them. Try to exercise for 30–40 minutes at least four days per week. Be sure to do warm-up stretching for 5 minutes before you begin, and cool down with stretching for 5 minutes after you've finished. Depending on your current level of physical conditioning, choose from the following list of exercises:

*Less Demanding*
- Golf: nine holes, no golf cart
- Active hatha yoga
- Tai chi, Chi Gong
- Swimming

*More Demanding*
- Vigorous walking: level ground, at least two miles
- Hiking: moderate pace, some hilly terrain
- Resistance training: light circuit training, or light hand weights (2–5 pounds) used with walking or hiking
- Pilates
- Tennis

*Meditate on This*

Meditation has been shown to have a positive effect on cortisol, one of the Warrior's main stress hormones. Excess cortisol is a major metabolic killer for the Warrior, increasing abdominal fat and interfering with proper memory function. Many people associate meditation with particular religious practices, but in reality it is just a state of concentrated attention. To meditate, one simply focuses their attention on some object or thought.

## How to Do It

There are excellent books available on meditation techniques, but here is a quick primer:

- Find a quiet place.
- Wear loose, comfortable clothing.
- Most people find that sitting cross-legged is naturally restful. You may want to sit on a cushion or towel. You can also use a chair, but try to sit only on the front half of the seat. Some people like to meditate with a blanket or shawl over their shoulders to prevent a chill.
- Your shoulders should be relaxed and your hands can rest in your lap.
- Keep your eyes "half open" without really looking at anything.
- Don't try to change the way you are breathing—just let your attention rest on the flow of your breath. The goal is to allow the "chattering" in your mind to fade away gradually.
- Relax every muscle in your body. Don't rush this, as it takes time to fully relax; relax bit by bit, starting at your toes and working up to your head.
- Visualize a place that calms you. It can be real or imaginary.

Meditation is an acquired skill, and Warriors benefit from a daily meditation practice. Start out with 5 minutes every day and gradually build up to 20 or 30 minutes. If you are like many other Warriors, you'll wear your daily meditation as a form of body armor and will soon refuse to leave home without doing your daily practice.

# CHAPTER TWENTY

# The Nomad Diet

**W**elcome, Nomad! This chapter contains all the information you'll need to get started on the diet for your GenoType. Diet buddy support, new research, and help with recipes and meal planning are only a mouse click away at the official GenoType Diet Web site (www.genotypediet.com).

The Nomad Diet is broken down by food category. Each category (Red Meats, Poultry, etc.) contains two lists. The list on the left contains the *Nomad Superfoods,* foods that act as medicines, balancing stress, regenerating genes, and repairing the digestive tract. Superfoods that help Nomads maintain ideal weight, increase muscle mass, and decrease body fat are identified by a diamond (◊) icon. To gain the maximum benefit from the Nomad Diet, these foods should be routinely consumed.

The list on the right contains the *Nomad Toxins,* foods that Nomad GenoTypes would be wise to avoid. Some foods on the Nomad Toxin list need only be avoided for a short period of time so that you can regain your balance. After 3–6 months, you can reintroduce these foods back into your diet in modest amounts. They are identified by a black dot (●) icon. Of course, if you are battling an illness or feel your weight beginning to creep back up, you may want to ramp up your compliance by avoiding these foods again for a while.

If a food is not listed, it is essentially *neutral,* meaning that the nutrients in it will benefit you but won't specifically help you restore balance to your genes or health to your cells. Feel free to eat these foods—but don't neglect the foods I recommend. A complete list of all the foods I've tested is available online (www.genotypediet.com).

## Red Meats

*Portion size: About the size of your hand (4–6 ounces)*
*Frequency: 2–3 times weekly; liver once weekly*

| Superfoods to Emphasize | Toxins to Limit or Avoid |
| --- | --- |
| Beef liver ◊ | Bear |
| Calf liver ◊ | Beef heart |
| Caribou | Boar |
| Goat | Ham |
| Kangaroo | Horse |
| Lamb | Pork |
| Moose | Pork, bacon |
| Mutton ◊ | Sweetbreads |
| Rabbit | |

## Poultry

*Portion size: About the size of your hand (4–6 ounces)*
*Frequency: 2–4 times weekly; liver once weekly*

| Superfoods to Emphasize | Toxins to Limit or Avoid |
| --- | --- |
| Emu ◊ | Chicken ● |
| Goose liver | Chicken liver |
| Ostrich ◊ | Cornish hen ● |
| Pheasant | Duck |
| Turkey ◊ | Duck liver |

| Superfoods to Emphasize | Toxins to Limit or Avoid |
|---|---|
| | Goose ● |
| | Grouse |
| | Guinea hen |
| | Partridge |
| | Quail |
| | Squab |

## Fish and Seafood

*Portion size: About the size of your hand (4–6 ounces)*
*Frequency: 4–5 times weekly*

| Superfoods to Emphasize | Toxins to Limit or Avoid |
|---|---|
| Bullhead | Anchovy ● |
| Catfish | Barracuda |
| Chub ◊ | Bass, blue gill |
| Cod ◊ | Bass, sea, lake |
| Croaker | Bass, striped |
| Cusk | Butterfish |
| Drum | Clam |
| Grouper ◊ | Conch |
| Hake ◊ | Crab |
| Halfmoon fish | Eel |
| Halibut ◊ | Frog |
| Harvest fish | Gray sole ● |
| Herring ◊ | Jellyfish |
| Mackerel, Atlantic ◊ | Lobster |
| Mackerel, Spanish ◊ | Mussels |
| Mahimahi | Octopus |
| Monkfish | Oyster |
| Mullet ◊ | Pollock, Atlantic |
| Muskellunge | Salmon, farm-raised |
| Ocean pout | Sea bream |
| Opaleye fish ◊ | Shrimp |
| Orange roughy ◊ | Skate |

| Superfoods to Emphasize | Toxins to Limit or Avoid |
| --- | --- |
| Parrotfish | Sole |
| Perch ◊ | Trout, farm-raised |
| Pickerel (walleye) | Trout, rainbow, wild ● |
| Pike | Trout, sea ● |
| Pilchard | Trout, steelhead, wild ● |
| Pompano | Turtle |
| Red snapper ◊ | Yellowtail |
| Rosefish ◊ | |
| Salmon, Alaskan ◊ | |
| Salmon, Atlantic, wild | |
| Salmon, chinook ◊ | |
| Salmon, sockeye ◊ | |
| Sardine ◊ | |
| Scallop | |
| Scrod ◊ | |
| Scup | |
| Shad | |
| Shark | |
| Sheepshead fish | |
| Smelt ◊ | |
| Squid, calamari | |
| Sturgeon | |
| Sucker | |
| Sunfish, pumpkinseed | |
| Swordfish | |
| Tilapia | |
| Tilefish ◊ | |
| Tuna, bluefin ◊ | |
| Tuna, skipjack | |
| Tuna, yellowfin | |
| Whitefish ◊ | |
| Whiting ◊ | |

# Eggs and Roes

*Portion size: 1 egg*
*Frequency: 5–7 servings per week*

| Superfoods to Emphasize | Toxins to Limit or Avoid |
|---|---|
| Carp roe ◊ | Egg, whole, duck |
| Caviar ◊ | Goose egg ● |
| Egg white, chicken | Quail egg |
| Egg white, duck | Salmon roe |
| Egg, whole, chicken ◊ | |
| Sailfish roe ◊ | |

# Dairy

*Portion size: Milk: 6 ounces; cheese: 2–4 ounces; ghee, butter, 1 tsp*
*Frequency: Cheeses: 4 times weekly*

| Superfoods to Emphasize | Toxins to Limit or Avoid |
|---|---|
| Brie cheese ◊ | American cheese |
| Camembert cheese ◊ | Blue cheese |
| Cheddar cheese ◊ | Casein |
| Colby cheese ◊ | Cheshire cheese ● |
| Edam cheese ◊ | Cottage cheese ● |
| Emmenthal, "Swiss" cheese ◊ | Cream cheese |
| Gouda cheese ◊ | Farmer cheese ● |
| Gruyère cheese ◊ | Feta cheese ● |
| Havarti cheese ◊ | Gorgonzola cheese |
| Jarlsberg cheese ◊ | Half-and-half |
| Kefir | Kefalotyri cheese |
| Manchego cheese ◊ | Limburger cheese |
| Muenster cheese | Milk, buttermilk, low-fat ● |
| Neufchâtel cheese | Milk, cow, skim or 2% |
| Parmesan cheese ◊ | Milk, cow, whole |
| Pecorino cheese ◊ | Milk, goat ● |
| Provolone cheese | Mozzarella cheese ● |
| Romanian Urdă | Paneer cheese ● |

| Superfoods to Emphasize | Toxins to Limit or Avoid |
| --- | --- |
| Romano cheese | Port du Salut cheese ● |
| Stilton cheese ◊ | Quark cheese ● |
| Yogurt | Ricotta cheese ● |
| | Roquefort cheese |
| | String cheese |
| | Whey protein powder |

## Vegetable Proteins

*Portion size: Nuts, seeds: ½ cup; nut butters: 1 tablespoon*
*Frequency: At least 5 times weekly*

| Superfoods to Emphasize | Toxins to Limit or Avoid |
| --- | --- |
| Almond ◊ | Adzuki bean |
| Almond butter ◊ | Black bean |
| Beechnut | Black-eyed pea |
| Brazil nut | Broad bean, fava |
| Butternut ◊ | Butter beans |
| Chestnut, Chinese | Cannellini bean ● |
| Chestnut, European | Carob ● |
| Chia seed, pinole | Cashew ● |
| Flaxseed, linseed ◊ | Cashew butter ● |
| Great northern bean | Copper bean ● |
| Green, string bean | Filbert, hazelnut |
| Hemp seed ◊ | Garbanzo bean, chickpea |
| Hickory nuts | Haricot beans |
| Macadamia ◊ | Kidney bean ● |
| Navy bean ◊ | Lentils, all types ● |
| Peas | Lentils, sprouted ● |
| Pecan ◊ | Lima bean |
| Snap bean | Lima bean flour |
| Soybean, sprouted | Lotus seeds |
| Tamarind bean | Lupin seeds |
| Walnut ◊ | Moth bean |
| Watermelon seeds ◊ | Mung bean |
| Yeast, nutritional ◊ | Natto |

| Superfoods to Emphasize | Toxins to Limit or Avoid |
|---|---|
|  | Peanut butter ● |
|  | Peanut flour ● |
|  | Peanuts ● |
|  | Pine nut, pignolia ● |
|  | Pistachio |
|  | Poppy seed |
|  | Pumpkin seed |
|  | Safflower seed |
|  | Sapodilla |
|  | Sesame butter, tahini |
|  | Sesame flour |
|  | Sesame seed |
|  | Soybean |
|  | Soybean flour |
|  | Soybean pasta |
|  | Soybean, tempeh |
|  | Soybean, tofu |
|  | Sunflower seed |
|  | Tempeh |
|  | White bean ● |
|  | Winged beans |
|  | Yard-long bean |

## Fats and Oils

*Portion size: 1 tablespoon*
*Frequency: 3–6 times weekly*

| Superfoods to Emphasize | Toxins to Limit or Avoid |
|---|---|
| Almond oil | Avocado oil |
| Babassu oil, extra virgin | Borage seed oil |
| Butter | Canola oil |
| Camelina oil ◊ | Coconut oil, commercial |
| Coconut oil, extra virgin ◊ | Corn oil |
| Cod liver oil ◊ | Cottonseed oil |

| Superfoods to Emphasize | Toxins to Limit or Avoid |
|---|---|
| Evening primrose oil | Lard |
| Flaxseed, linseed oil ◊ | Margarine |
| Ghee (clarified butter) | Palm oil |
| Hazelnut nut oil | Peanut oil |
| Hemp seed oil ◊ | Pumpkin seed oil |
| Herring oil ◊ | Safflower oil |
| Macadamia oil ◊ | Sesame oil |
| Olive oil ◊ | Soybean oil |
| Perilla seed oil ◊ | Sunflower oil |
| Rice bran oil | Wheat germ oil |
| Salmon oil ◊ | |
| Walnut oil | |

## Carbohydrates

*Portion size: ½ cup grains, cereals, rice; ½ muffin; 1 slice bread*
*Frequency: 2–3 times daily*

| Superfoods to Emphasize | Toxins to Limit or Avoid |
|---|---|
| Flaxseed bread | Amaranth |
| Fonio | Artichoke flour/pasta |
| Gluten-free and corn-free breads | Barley |
| Job's tears | Buckwheat, kasha, soba |
| Kudzu | Cornmeal, hominy, polenta |
| Larch fiber | Essene, Manna bread ● |
| Millet | Lentil flour, dhal |
| Oat bran | Mastic gum |
| Oat flour | Poppadom ● |
| Quinoa | Rice, wild ● |
| Rice bran | Rye ● |
| Rice flour, brown | Rye flour ● |
| Rice flour, white | Sorghum |
| Rice, basmati | Tapioca, manioc, cassava |
| Rice, brown | Teff |
| Rice, white | Wheat, bran, germ |
| Taro | Wheat, bulgur |

| Superfoods to Emphasize | Toxins to Limit or Avoid |
|---|---|
| | Wheat, durum, semolina |
| | Wheat, emmer, einkorn ● |
| | Wheat, kamut |
| | Wheat, spelt ● |
| | Wheat, white flour |
| | Wheat, whole grain |
| | Wheat, 100% sprouted ● |

## Live Foods

*Portion size: 1 cup*
*Frequency: At least 4–5 servings daily*

| Superfoods to Emphasize | Toxins to Limit or Avoid |
|---|---|
| Alfalfa sprouts | Aloe vera |
| Asparagus ◊ | Artichoke |
| Asparagus pea | Avocado |
| Beet ◊ | Chicory ● |
| Beet greens | Collard greens ● |
| Bok choy, pak choi | Daikon radish ● |
| Broccoflower | Fenugreek |
| Broccoli | Jerusalem artichoke |
| Broccoli rabe | Jicama ● |
| Brussels sprout | Kohlrabi ● |
| Cabbage ◊ | Corn, popcorn |
| Carrot ◊ | Okra ● |
| Cauliflower ◊ | Olive, black |
| Celeriac | Olive, green ● |
| Celery ◊ | Parsnip ● |
| Chayote, pipinella | Pickle, brine |
| Chinese kale, kai-lan | Pickle, vinegar |
| Cucumber ◊ | Potato, white with skin ● |
| Eggplant ◊ | Pumpkin |
| Ginger ◊ | Quorn |
| Grape leaves | Radish ● |

| Superfoods to Emphasize | Toxins to Limit or Avoid |
|---|---|
| Hearts of palm | Radish sprouts ● |
| Horseradish | Rhubarb |
| Jute, potherb | Scallion ● |
| Kale | Tomatillo |
| Leek | Tomato |
| Lettuce, green leaf, iceberg | Turnip ● |
| Lettuce, romaine | Water chestnut |
| Mushroom, commercial ◊ | |
| Mushroom, cremini ◊ | |
| Mushroom, enoki ◊ | |
| Mushroom, maitake ◊ | |
| Mushroom, oyster ◊ | |
| Mushroom, portobello ◊ | |
| Mushroom, shiitake | |
| Mushroom, straw ◊ | |
| Mustard greens | |
| Onion, all types ◊ | |
| Pepper, bell | |
| Pepper, chili, jalapeño | |
| Rutabaga ◊ | |
| Sauerkraut ◊ | |
| Sea cucumber | |
| Sea vegetables, kelp | |
| Sea vegetables, spirulina | |
| Sea vegetables, wakame | |
| Spinach ◊ | |
| Squash | |
| Sweet potato ◊ | |
| Swiss chard | |
| Turnip greens | |
| Watercress | |
| Yam | |
| Zucchini | |

# Fruits

*Portion size: 1 cup fruit or 1 medium-size fruit*
*Frequency: At least 3 servings daily*

| Superfoods to Emphasize | Toxins to Limit or Avoid |
| --- | --- |
| Apple | Banana ● |
| Apricot | Bitter melon |
| Blueberry ◊ | Cherimoya ● |
| Breadfruit | Coconut meat ● |
| Canistel | Currants ● |
| Cantaloupe melon ◊ | Dewberry |
| Cherry ◊ | Guava |
| Cranberry ◊ | Huckleberry |
| Date | Jackfruit ● |
| Durian | Kumquat ● |
| Elderberry | Loquat |
| Fig | Mango |
| Grape ◊ | Noni ● |
| Grapefruit | Orange |
| Kiwi ◊ | Persimmon |
| Lemon | Plantain ● |
| Lime ◊ | Pomegranate |
| Lingonberry | Prickly pear |
| Loganberry | Quince |
| Mamey sapote | Sago palm |
| Musk melon ◊ | Star fruit, carambola |
| Nectarine ◊ | Tamarillo ● |
| Papaya | |
| Passion fruit | |
| Pawpaw ◊ | |
| Peach ◊ | |
| Pear ◊ | |
| Persian melon | |
| Pineapple | |
| Plum | |
| Prune | |
| Raisin | |
| Raspberry ◊ | |

| Superfoods to Emphasize | Toxins to Limit or Avoid |
| --- | --- |
| Spanish melon | |
| Strawberry ◊ | |
| Tangerine | |
| Watermelon ◊ | |

## Spices

*Portion size: 1 teaspoon*
*Frequency: At least 1–2 servings daily*

| Superfoods to Emphasize | Toxins to Limit or Avoid |
| --- | --- |
| Basil ◊ | Allspice ● |
| Bay leaf | Anise ● |
| Chives | Caper ● |
| Cilantro ◊ | Caraway ● |
| Curry | Celery seed ● |
| Garlic | Chocolate ● |
| Lemon grass, citronella | Cinnamon ● |
| Licorice root | Guarana |
| Nutmeg | Mustard, dry ● |
| Oregano | Pepper, black ● |
| Paprika | Pepper, red flakes ● |
| Parsley ◊ | |
| Peppermint | |
| Rosemary | |
| Sage | |
| Thyme ◊ | |
| Turmeric | |

# Beverages

*Portion size: 6–8-ounces*
*Frequency: 2–4 servings daily*

| Superfoods to Emphasize | Toxins to Limit or Avoid |
|---|---|
| Beer ◊ | Beet juice |
| Blackberry juice | Coffee ● |
| Carrot juice | Cola beverages |
| Celery juice ◊ | Diet sodas |
| Cherry juice | Kombucha tea ● |
| Cranberry juice | Liquor, distilled |
| Elderberry juice | Milk, almond ● |
| Grape juice | Milk, coconut |
| Grapefruit juice | Milk, soy |
| Lemon and water | Orange juice |
| Milk, rice | Pomegranate juice |
| Pineapple juice | Tea, black |
| Tea, gingerroot | Tomato juice |
| Tea, ginseng ◊ | Wine, white ● |
| Tea, green, kukicha, bancha ◊ | |
| Tea, licorice root ◊ | |
| Tea, verbena ◊ | |
| Watermelon juice ◊ | |
| Wine, red | |

# Condiments and Additives

*Portion size: 1 teaspoon to 1 tablespoon*
*Frequency: Use as needed*

| Superfoods to Emphasize | Toxins to Limit or Avoid |
|---|---|
| Agave syrup ◊ | Acacia (gum arabic) |
| Epazote | Arrowroot ● |
| Fruit pectin | Aspartame |
| Honey ◊ | Barley malt |
| Maple syrup | Calcium diglutamate |

| Superfoods to Emphasize | Toxins to Limit or Avoid |
|---|---|
| Molasses ◊ | Carrageenan |
| Molasses, blackstrap ◊ | Cornstarch |
| Rice syrup | Cream of tartar |
| Roselle | Gelatin, plain |
| Sea salt | Glutamic acid |
| Vegetable glycerine ◊ | Guar gum |
| Yeast extract spread, Marmite | High-fructose corn syrup |
| Yeast, nutritional ◊ | Ketchup |
| | Monoammonium glutamate |
| | Monopotassium glutamate |
| | MSG (monosodium glutamate) |
| | Mayonnaise ● |
| | Mayonnaise, tofu |
| | Miso |
| | Mustard, with vinegar ● |
| | Pickle relish |
| | Sea vegetables, agar |
| | Sea vegetables, Irish moss |
| | Soybean sauce, tamari |
| | Sugar, brown, white |
| | Stevia |
| | Umeboshi plum, vinegar ● |
| | Vinegar, white |
| | Vinegar, wine ● |
| | Worcestershire sauce |

# GenoType Nomad Supplement Guide

These supplements can help improve your results while on the Nomad Diet. Most are readily available in health food stores, while a few are a bit more obscure. Note that these supplements are solely recommended for use as part of the Nomad GenoType Diet program. To examine how other supplements you may be using rate with GenoType Nomad, or for more detailed information on the science behind the use of this supplement protocol, visit the official GenoType Diet Web site online at www.genotypediet.com. Remember to discuss the use of any nutritional supplements with your physician before embarking on a supplement program.

*Supplements that balance the Nomad's production of nitric oxide, improving their cardiovascular, neurological, and detoxification functions:*
- l-Arginine. Typical daily dose: 250–500 mg
- *Cordyceps sinensis.* Typical daily dose: 200–500 mg
- Ginseng (*Panax* species). Typical daily dose: 50–250 mg

*Supplements that enhance the Nomad's internal ecosystem, slowing down premature aging and enhancing immune function:*
- Larch (*Larix* species) Arabinogalactan. Typical daily dose: 250–400 mg
- Vitamin B$_7$ (Biotin). Typical daily dose: 1–4 mg
- Vitamin B$_9$ (Folic Acid). Typical daily dose: 400–800 mcg
- *Schizandra chinensis.* Typical daily dose: 300–500 mg

*Supplements that improve the receptor sensitivity in Nomads, improving their response to stress, increasing sensitivity to the body's own hormones, and improving the metabolic functions:*
- Forskolin extract (*Coleus* species). Typical daily dose: 150–250 mg
- l-Lysine. Typical daily dose: 250–500 mg
- Creatine powder. Typical daily dose: 1–3 grams

# GenoType Nomad Lifestyle
## Meal Planning

The Nomad has a range of food combinations and cuisines available for imaginative meal planning. We've collected a variety of meal plans (even plans for families with different member GenoTypes) and hundreds of tasty recipes for each GenoType online at www.genotypediet.com.

## Exercise Guide

The Nomad exercise protocol is designed to increase muscle growth, enhance the metabolic functions, and reduce stress. Nomad GenoTypes do best with stretching and lengthening exercises that also have a resistance component to them. Try to exercise for 30–40 minutes at least 5 days per week, using a combination of the *More* and *Less Demanding* exercises, depending on your current level of physical conditioning. Be sure to do warm-up stretching for 5 minutes before you begin, and cool down with stretching for 5 minutes after you've finished. Choose from the following list of exercises:

*Less Demanding*
- Golf: nine holes, no golf cart
- Active hatha yoga
- Tai chi, Chi Gong
- Swimming
- Pilates
- Vigorous walking: level ground, at least 2 miles
- Hiking: moderate pace, some hilly terrain

*More Demanding*
- Resistance training: light circuit training, or light hand weights (2–5 pounds) used with walking or hiking
- Intense competitive sports (martial arts, basketball, soccer)
- Jogging
- Tennis

*Brushing Up on Things*
The skin has more immune and nervous system tissue than any other organ of the body. Dry brushing has a wonderfully balancing effect on Nomads. It brings blood to their skin, tonifies their nervous system, and helps enhance their antiviral defenses. Try doing this practice twice weekly for a month and you'll soon be hooked!

## How to Do It

- To dry brush, use a soft natural-fiber brush with a long handle so you are able to reach all areas of your body. A loofah (luffa) sponge or a rough towel can also be used.
- Stand in the shower with the water off.
- Starting at your feet, start brushing in small circles toward your heart. Apply very light pressure to those areas where the skin is thin and deeper pressure on places like the soles of the feet.
- Avoid broken skin, skin rashes, or areas where the skin is thin, such as the face or inner thighs.
- After you've finished both legs, move on to your arms. Always brush from your fingertips toward your heart.
- Reach around and brush from your back toward your stomach.
- When you are finished, begin showering with hot water for 5 minutes, then end with cold water.
- Apply a light coating of an acceptable oil (babsassu oil, extra-virgin coconut oil, and almond oil are good choices) and massage in well.

# The Future Beyond Tomorrow

O ver three decades ago, I was a typically mixed-up teenager attending a highly regarded Catholic high school. During sophomore year, we had the good fortune to have the wizened Father Kenney as our religion teacher. His lectures were always well attended, perhaps as the result of his policy of allowing students to smoke cigarettes in class.

Between long puffs of his own, followed by streaming eruptions of grayish-blue smoke, Father Kenney would hold forth on the universe.

"Father Kenney, why do bad things happen to good people?"

"Well, because God so values free choice, he is willing to let events, both good and bad, unfold."

"How can you accept science and evolution, and still believe in God?"

Long inhale. Slow exhale.

"Because there is nothing so small, or so elemental, that you cannot see the hand of God in it."

Now, a lot of evolutionary scientists are atheists. I happen to not be one of them. However, the problem with mixing religion and science is that you are trying to reconcile certainty and uncertainty, and if you are certain about something, your thoughts about it can't grow very much.

Genetics is intimately linked with evolution and natural selection. When genes reproduce, the process occasionally screws up and a mistake, or *mutation*, occurs. Most mutations are bad, but every once in a while a mutation occurs that alters the odds of that species surviving. If it is a big enough improvement, it enters the gene pool and perhaps a new species is born. Theoretically, this all happens in a random (if somewhat heartless) manner and over a very long time. Talking about genes is always a bit risky—many people admit to being concerned or scared by the whole thing and perhaps wish that geneticists would just leave well enough alone. However, genetics is rich with possibilities despite the fact that what is revealed often challenges long-accepted ideas and notions.

When I was five years old, I was stricken with a bad case of the measles. To help lift my spirits up, my parents brought home a new toy. It was a terrarium type of tank to which you added a special powder and water. Over the next few days, my listlessness and apathy turned to won-der as brightly colored stalagmites materialized and grew ever higher. Each morning upon awakening, I would rush to my desk across the room to marvel at what magic had happened while I was asleep.

But what would have been the response if my parents had just given me a premade stalactite garden? I can tell you. About two minutes of attention and then relegation to the closet. What was fascinating about the toy was its *process*, not its *outcome*. If we assume that events that fas-cinate God are similar to the types of things that fascinate a little boy, then it is possible to have faith and also live in a fact-based world.

Epigenetics takes the evolution discussion to the next level. Rather than worrying about whether you came from an amoeba or some small mammal that made a wrong turn, you can worry about things much more relevant—like whether the weaknesses introduced into your heredity from grandparents and parents will come home to roost within you.

More important, you can do more than just worry—you can do some-thing about it. By putting the GenoType Diet to work, not only do you make choices that will change your destiny, but under the right conditions they'll go on to change the destiny of your descendants.

There is a famous story about a rabbi and a young child. The rabbi was planting a plum tree sapling. Like most kids, this one was direct and to the point. "Isn't it kind of stupid to plant a tiny sapling when at your age there is no chance that you will be alive to taste those plums?" said the child. The rabbi thought a minute and answered the child's question with one of his own.

"Do you like plums?"

Hopefully, you get the point. There is a future somewhere out there. If we care to, we can live in the present and let the future take care of itself. However, for those of us with the ambition and who are ready to do the work, epigenetics holds the promise of an improvable future of health for us and those who come after us.

What if I told you that in four generations you and your immediate descendants could change the epigenetic patterns of inheritance in your family line and eliminate diabetes, heart disease, and some forms of cancer? We all know a family of long-lived people or a family where nobody seems to get cancer but everyone worries about getting Alzheimer's disease, which seems to strike half of all the family members. These are epigenetic traits within those families. Carried over several generations, they imprint themselves onto the family's epigenome. Drugs, toxins, and poor diet all work their way into your family's epigenome as well. That's why we see so much attention deficit disorder (ADD), obesity, high blood pressure, cancer, and diabetes in our modern population.

There is much to learn from the golden Agouti mouse. Think about it. If we can imprint bad stuff through lack of attention or ignorance, can we, with knowledge, just as easily imprint good things as well?

Your grandchildren could be the first to parent this new generation of epigenetically healthy children. All it takes is guts, vision, and a plan.

That's what the GenoType Diet is all about.

# The Advanced GenoType Calculator Tables

This appendix contains the four jump tables that are used to calculate GenoType with the advanced biomarker secretor status. Refer to the section "Advanced GenoType Testing" in Chapter 6 for more information.

| And . . . | Blood Type | Rhesus (Rh) | Secretor Status | GenoType |
|---|---|---|---|---|
| Your index fingers are longer than your ring fingers on both hands. | A | + | Secretor | GT3 TEACHER |
| | | | Non-secretor | GT3 TEACHER (women) GT4 EXPLORER (men) |
| | | − | Secretor | GT3 TEACHER (women) GT4 EXPLORER (men) |
| | | | Non-secretor | GT4 EXPLORER |
| | AB | + | Secretor | GT6 NOMAD |
| | | | Non-secretor | GT4 EXPLORER |
| | | − | Secretor | GT4 EXPLORER (men) GT6 NOMAD (women) |
| | | | Non-secretor | GT4 EXPLORER |
| | B | + | Secretor | GT6 NOMAD |
| | | | Non-secretor | GT6 NOMAD |
| | | − | Secretor | GT4 EXPLORER (men) GT6 NOMAD (women) |
| | | | Non-secretor | GT4 EXPLORER |
| | O | + | Secretor | GT2 GATHERER |
| | | | Non-secretor | GT2 GATHERER |
| | | − | Secretor | GT2 GATHERER |
| | | | Non-secretor | GT2 GATHERER (women) GT4 EXPLORER (men) |
| Your ring fingers are longer on both hands. | A | + | Secretor | GT3 TEACHER |
| | | | Non-secretor | GT3 TEACHER (men) GT4 EXPLORER (women) |
| | | − | Secretor | GT3 TEACHER |
| | | | Non-secretor | GT4 EXPLORER |
| | AB | + | Secretor | GT6 NOMAD |
| | | | Non-secretor | GT6 NOMAD |
| | | − | Secretor | GT6 NOMAD (men) GT4 EXPLORER (women) |
| | | | Non-secretor | GT4 EXPLORER |

**Advanced GenoType Calculator: Table 1 (continued)**

Use this table if:

- Your TORSO is LONGER than or equal in length to your LEGS
- Your UPPER LEG is LONGER than your LOWER LEG

| And ... | Blood Type | Rhesus (Rh) | Secretor Status | GenoType |
|---------|-----------|-------------|-----------------|----------|
| Your ring fingers are longer on both hands. | B | + | Secretor | GT6 NOMAD |
| | | | Non-secretor | GT6 NOMAD |
| | | − | Secretor | GT4 EXPLORER (women) GT6 NOMAD (men) |
| | | | Non-secretor | GT4 EXPLORER |
| | O | + | Secretor | GT1 HUNTER |
| | | | Non-secretor | GT1 HUNTER (men) GT4 EXPLORER (women) |
| | | − | Secretor | GT1 HUNTER |
| | | | Non-secretor | GT1 HUNTER (men) GT4 EXPLORER (women) |
| Your index finger is longer on one hand and your ring finger is longer on the other. | A | + | Secretor | GT3 TEACHER |
| | | | Non-secretor | GT3 TEACHER |
| | | − | Secretor | GT3 TEACHER |
| | | | Non-secretor | GT3 TEACHER |
| | AB | + | Secretor | GT5 WARRIOR |
| | | | Non-secretor | GT5 WARRIOR |
| | | − | Secretor | GT4 EXPLORER |
| | | | Non-secretor | GT4 EXPLORER |
| | B | + | Secretor | GT2 GATHERER |
| | | | Non-secretor | GT2 GATHERER |
| | | − | Secretor | GT4 EXPLORER |
| | | | Non-secretor | GT4 EXPLORER |
| | O | + | Secretor | GT2 GATHERER |
| | | | Non-secretor | GT2 GATHERER (women) GT4 EXPLORER (men) |
| | | − | Secretor | GT2 GATHERER (women) GT4 EXPLORER (men) |
| | | | Non-secretor | GT4 EXPLORER |

**Advanced GenoType Calculator: Table 2**

Use this table if:

- Your TORSO is LONGER than or equal in length to your LEGS
- Your LOWER LEG is LONGER than or equal to your UPPER LEG

| And . . . | Blood Type | Rhesus (Rh) | Secretor Status | GenoType |
|---|---|---|---|---|
| Your index fingers are longer than your ring fingers on both hands. | A | + | Secretor | GT5 WARRIOR |
| | | | Non-secretor | GT5 WARRIOR |
| | | − | Secretor | GT5 WARRIOR |
| | | | Non-secretor | GT4 EXPLORER |
| | AB | + | Secretor | GT5 WARRIOR (men) GT6 NOMAD (women) |
| | | | Non-secretor | GT5 WARRIOR (men) GT6 NOMAD (women) |
| | | − | Secretor | GT5 WARRIOR (men) GT6 NOMAD (women) |
| | | | Non-secretor | GT5 WARRIOR (men) GT6 NOMAD (women) |
| | B | + | Secretor | GT6 NOMAD |
| | | | Non-secretor | GT2 GATHERER |
| | | − | Secretor | GT2 GATHERER |
| | | | Non-secretor | GT2 GATHERER |
| | O | + | Secretor | GT2 GATHERER |
| | | | Non-secretor | GT2 GATHERER |
| | | − | Secretor | GT2 GATHERER |
| | | | Non-secretor | GT2 GATHERER (women) GT4 EXPLORER (men) |
| Your ring fingers are longer than your index fingers on both hands. | A | + | Secretor | GT3 TEACHER |
| | | | Non-secretor | GT3 TEACHER |
| | | − | Secretor | GT3 TEACHER (men) GT4 EXPLORER (women) |
| | | | Non-secretor | GT3 TEACHER (men) GT4 EXPLORER (women) |
| | AB | + | Secretor | GT3 TEACHER |
| | | | Non-secretor | GT3 TEACHER |
| | | − | Secretor | GT3 TEACHER |
| | | | Non-secretor | GT4 EXPLORER |

*Advanced GenoType Calculator: Table 2 (continued)*

*Use this table if:*

- Your TORSO is LONGER than or equal in length to your LEGS
- Your LOWER LEG is LONGER than or equal to your UPPER LEG

| And . . . | Blood Type | Rhesus (Rh) | Secretor Status | GenoType |
|---|---|---|---|---|
| Your ring fingers are longer than your index fingers on both hands. | B | + | Secretor | GT6 NOMAD |
| | | | Non-secretor | GT6 NOMAD *(men)* GT4 EXPLORER *(women)* |
| | | – | Secretor | GT4 EXPLORER |
| | | | Non-secretor | GT4 EXPLORER |
| | O | + | Secretor | GT1 HUNTER |
| | | | Non-secretor | GT1 HUNTER |
| | | – | Secretor | GT4 EXPLORER |
| | | | Non-secretor | GT1 HUNTER *(men)* GT4 EXPLORER *(women)* |
| Your index finger is longer on one hand and your ring finger is longer on the other. | A | + | Secretor | GT3 TEACHER |
| | | | Non-secretor | GT3 TEACHER |
| | | – | Secretor | GT3 TEACHER |
| | | | Non-secretor | GT3 TEACHER |
| | AB | + | Secretor | GT3 TEACHER |
| | | | Non-secretor | GT3 TEACHER |
| | | – | Secretor | GT3 TEACHER |
| | | | Non-secretor | GT4 EXPLORER |
| | B | + | Secretor | GT6 NOMAD |
| | | | Non-secretor | GT2 GATHERER |
| | | – | Secretor | GT4 EXPLORER |
| | | | Non-secretor | GT4 EXPLORER |
| | O | + | Secretor | GT2 GATHERER |
| | | | Non-secretor | GT2 GATHERER |
| | | – | Secretor | GT2 GATHERER |
| | | | Non-secretor | GT4 EXPLORER |

Advanced GenoType Calculator: Table 3

Use this table if:

- Your LEGS are LONGER than or equal in length to your TORSO
- Your UPPER LEG is LONGER than your LOWER LEG

| And . . . | Blood Type | Rhesus (Rh) | Secretor Status | GenoType |
|---|---|---|---|---|
| Your index fingers are longer than your ring fingers on both hands. | A | + | Secretor | GT5 WARRIOR |
| | | | Non-secretor | GT5 WARRIOR |
| | | − | Secretor | GT4 EXPLORER |
| | | | Non-secretor | GT4 EXPLORER |
| | AB | + | Secretor | GT5 WARRIOR (men) GT6 NOMAD (women) |
| | | | Non-secretor | GT5 WARRIOR (men) GT6 NOMAD (women) |
| | | − | Secretor | GT5 WARRIOR |
| | | | Non-secretor | GT5 WARRIOR |
| | B | + | Secretor | GT2 GATHERER (men) GT6 NOMAD (women) |
| | | | Non-secretor | GT2 GATHERER |
| | | − | Secretor | GT4 EXPLORER (men) GT6 NOMAD (women) |
| | | | Non-secretor | GT4 EXPLORER |
| | O | + | Secretor | GT2 GATHERER |
| | | | Non-secretor | GT2 GATHERER (women) GT4 EXPLORER (men) |
| | | − | Secretor | GT2 GATHERER |
| | | | Non-secretor | GT2 GATHERER (women) GT4 EXPLORER (men) |
| Your ring fingers are longer than your index fingers on both hands. | A | + | Secretor | GT5 WARRIOR |
| | | | Non-secretor | GT4 EXPLORER (men) GT5 WARRIOR (women) |
| | | − | Secretor | GT5 WARRIOR |
| | | | Non-secretor | GT4 EXPLORER |
| | AB | + | Secretor | GT5 WARRIOR |
| | | | Non-secretor | GT5 WARRIOR |
| | | − | Secretor | GT5 WARRIOR |
| | | | Non-secretor | GT4 EXPLORER |

*Advanced GenoType Calculator: Table 3 (continued)*

*Use this table if:*

- Your LEGS are LONGER than or equal in length to your TORSO
- Your UPPER LEG is LONGER than your LOWER LEG

| And . . . | Blood Type | Rhesus (Rh) | Secretor Status | GenoType |
|---|---|---|---|---|
| Your ring fingers are longer than your index fingers on both hands. | B | + | Secretor | GT6 NOMAD |
| | | | Non-secretor | GT6 NOMAD |
| | | – | Secretor | GT4 EXPLORER *(women)* GT6 NOMAD *(men)* |
| | | | Non-secretor | GT4 EXPLORER |
| | O | + | Secretor | GT1 HUNTER |
| | | | Non-secretor | GT1 HUNTER |
| | | – | Secretor | GT1 HUNTER *(men)* GT4 EXPLORER *(women)* |
| | | | Non-secretor | GT1 HUNTER *(men)* GT4 EXPLORER *(women)* |
| Your index finger is longer on one hand and your ring finger is longer on the other. | A | + | Secretor | GT3 TEACHER |
| | | | Non-secretor | GT3 TEACHER |
| | | – | Secretor | GT3 TEACHER |
| | | | Non-secretor | GT4 EXPLORER |
| | AB | + | Secretor | GT3 TEACHER |
| | | | Non-secretor | GT4 EXPLORER |
| | | – | Secretor | GT3 TEACHER |
| | | | Non-secretor | GT4 EXPLORER |
| | B | + | Secretor | GT2 GATHERER |
| | | | Non-secretor | GT2 GATHERER |
| | | – | Secretor | GT4 EXPLORER |
| | | | Non-secretor | GT4 EXPLORER |
| | O | + | Secretor | GT1 HUNTER |
| | | | Non-secretor | GT2 GATHERER |
| | | – | Secretor | GT1 HUNTER |
| | | | Non-secretor | GT2 GATHERER |

**Advanced GenoType Calculator: Table 4**

**Use this table if:**
- Your LEGS are LONGER than your TORSO
- Your LOWER LEG is LONGER than or equal to your UPPER LEG

| And . . . | Blood Type | Rhesus (Rh) | Secretor Status | GenoType |
|---|---|---|---|---|
| Your index fingers are longer than your ring fingers on both hands. | A | + | Secretor | GT5 WARRIOR |
| | | | Non-secretor | GT5 WARRIOR |
| | | − | Secretor | GT5 WARRIOR |
| | | | Non-secretor | GT4 EXPLORER (men) GT5 WARRIOR (women) |
| | AB | + | Secretor | GT5 WARRIOR (men) GT6 NOMAD (women) |
| | | | Non-secretor | GT5 WARRIOR (men) GT6 NOMAD (women) |
| | | − | Secretor | GT5 WARRIOR |
| | | | Non-secretor | GT5 WARRIOR |
| | B | + | Secretor | GT6 NOMAD |
| | | | Non-secretor | GT2 GATHERER (men) GT6 NOMAD (women) |
| | | − | Secretor | GT2 GATHERER |
| | | | Non-secretor | GT2 GATHERER |
| | O | + | Secretor | GT2 GATHERER |
| | | | Non-secretor | GT2 GATHERER (women) GT4 EXPLORER (men) |
| | | − | Secretor | GT2 GATHERER (women) GT4 EXPLORER (men) |
| | | | Non-secretor | GT4 EXPLORER |
| Your ring fingers are longer than your index fingers on both hands. | A | + | Secretor | GT3 TEACHER (men) GT5 WARRIOR (women) |
| | | | Non-secretor | GT3 TEACHER (men) GT5 WARRIOR (women) |
| | | − | Secretor | GT3 TEACHER (men) GT5 WARRIOR (women) |
| | | | Non-secretor | GT3 TEACHER (men) GT4 EXPLORER (women) |
| | AB | + | Secretor | GT5 WARRIOR (women) GT6 NOMAD (men) |
| | | | Non-secretor | GT5 WARRIOR (women) GT6 NOMAD (men) |

*Advanced GenoType Calculator: Table 4 (continued)*

Use this table if:

- Your LEGS are LONGER than your TORSO
- Your LOWER LEG is LONGER than or equal to your UPPER LEG

| And . . . | Blood Type | Rhesus (Rh) | Secretor Status | GenoType |
|---|---|---|---|---|
| Your ring fingers are longer than your index fingers on both hands. | AB | − | Secretor | GT5 WARRIOR |
| | | | Non-secretor | GT5 WARRIOR |
| | B | + | Secretor | GT6 NOMAD |
| | | | Non-secretor | GT6 NOMAD (men) GT4 EXPLORER (women) |
| | | − | Secretor | GT6 NOMAD (men) GT4 EXPLORER (women) |
| | | | Non-secretor | GT6 NOMAD (men) GT4 EXPLORER (women) |
| | O | + | Secretor | GT1 HUNTER |
| | | | Non-secretor | GT1 HUNTER |
| | | − | Secretor | GT1 HUNTER |
| | | | Non-secretor | GT1 HUNTER |
| Your index finger is longer on one hand and your ring finger is longer on the other. | A | + | Secretor | GT3 TEACHER |
| | | | Non-secretor | GT5 WARRIOR |
| | | − | Secretor | GT3 TEACHER |
| | | | Non-secretor | GT5 WARRIOR |
| | AB | + | Secretor | GT5 WARRIOR (men) GT6 NOMAD (women) |
| | | | Non-secretor | GT5 WARRIOR (men) GT6 NOMAD (women) |
| | | − | Secretor | GT5 WARRIOR |
| | | | Non-secretor | GT5 WARRIOR |
| | B | + | Secretor | GT6 NOMAD |
| | | | Non-secretor | GT6 NOMAD (women) GT4 EXPLORER (men) |
| | | − | Secretor | GT4 EXPLORER |
| | | | Non-secretor | GT4 EXPLORER |
| | O | + | Secretor | GT1 HUNTER |
| | | | Non-secretor | GT1 HUNTER |
| | | − | Secretor | GT1 HUNTER |
| | | | Non-secretor | GT4 EXPLORER |

# GenoType Diet Terms

**Acetylator:** A genetic polymorphism (see page 301) that determines how fast an individual can detoxify many substances, including caffeine. Slow acetylators, such as the Explorer GenoType, are often drug-sensitive, whereas fast acetylators, such as the Warrior GenoType, often have problems removing certain carcinogens.

**Advanced Glycation End-Product:** Advanced glycation end-products (AGEs) are molecules made of carbohydrates bound to a protein (glycoprotein). Unlike most glycoproteins made by the body that are a result of enzyme activity, AGEs are a result of metabolic misfortune, a type of "browning reaction" that is essentially not reversible. Many of the manifestations of aging are the result of the deposit of AGEs in organs and tissues.

**Alleles:** The "alternate" set (or pair) of genes received from both the father and the mother. In most circumstances, one allele for a particular trait will be dominant or recessive to another.

**Andric:** Male- or masculine-shaped because of the presence of more androgen (male) hormone stimulation in utero, regardless of whether the individual is male or female. The shape is longer, leaner, and more muscular. A characteristic of the andric body type is a wide opening of the space between the legs above the knees.

**Archetype:** A generic, idealized model of a person from which similar instances are derived, copied, patterned, or emulated. The concept of a GenoType (see below) as defined in this book is a portmanteau of "Genetic Archetype."

**Biometrics:** Biometrics is literally "the measure of living things." It's a way of measuring your bone lengths and other key elements of your physical self. Fingerprint analysis, also called dermatoglyphics, is a subset of biometrics as well.

**Epigenetic:** Refers to the interaction between your genes and the environment, resulting in a change in the expression of the genetic material, although the DNA has not been altered. Epigenetics is sometimes called "post-genomic inheritance." It is the study of how our genes respond to the environment, creating differences that we can pass along to our children.

**Genes:** A gene is a discrete section of DNA that carries all the information needed to convey a specific trait. There are approximately 25,000 to 30,000 genes in a human.

**GenoType:** In conventional genetics, *genotype* is the genetic basis of a trait, as compared to *phenotype,* which is the physical appearance that the genotype trait produces. In this book, the term GenoType is a portmanteau of "Genetic Archetype," a series of strengths and weaknesses that result from the interaction of the genetic, epigenetic, and environmental influences on a person, especially during their prenatal and early-childhood periods and how they organize themselves into recognizable biochemical profiles over the course of the individual's life.

**Glycation:** Glycation occurs when a sugar molecule, such as fructose, from fruit or glucose, from refined grains, binds to a protein and damages it. The damaged protein interferes with organ function, blood profusion, hormone receptivity, and kidney function, and it might cause cataracts and neuron damage as well. See also **Advanced Glycation End-Product.**

**Gynic:** Female- or feminine-shaped because of the presence of more female hormones in utero, although the individual can be either male or female. The shape is more rounded and soft. A characteristic of the gynic body type is a narrow opening of the space between the legs above the knees.

**Histone:** Spool-like molecules that cause DNA to wind into dense coils. Coiling up DNA silences genes. Uncoiling histones causes the DNA to become available for reading and thereby activates the genes.

**Lectins:** Lectins are proteins found in food that interact with many of the sugars found on the exterior of our bodies' cells. Many lectins are specific for a particular blood type; when you eat a food containing incompatible lectins, it can interfere with digestion, metabolism, and proper immune system functioning.

**Methylation:** One of the prime mechanisms of epigenetic gene regulation, methylation refers to the attachment of a methyl group to the DNA molecule in such a way as to prevent the gene from being "read." Such genes are considered "silenced."

**Polymorphism:** Literally meaning "many shapes," polymorphisms are multiple alleles of a gene within a population, usually expressing different phenotypes. See **GenoType.**

**Somatotype:** Basic classifications of body types developed in the 1940s by American psychologist William Sheldon, according to the prominence of different basic tissue types. These roughly translate to the rounder-shaped *endomorph,* the lanky *ectomorph,* and the muscular *mesomorph.*

**Symmetry:** Symmetry is the measurable similarity between the sides of the body. It is considered a subtle indicator of fitness, since asymmetries between the left and right sides of the body often indicate environmental instability during fetal development.

**White Lines:** Appearing throughout an otherwise inked fingerprint pattern, white lines are a sign that the height of the fingerprint ridges is low, to the point where underlying creases of the skin show through. White lines throughout the fingerprints are an indication that the individual may have gluten and/or lectin intolerance.

**Worldview:** Refers to the framework of actions and reactions through which an individual GenoType interprets the world and interacts in it. The six GenoTypes share three basic worldviews: the Reactive worldview (Hunters and Explorers); the Tolerant worldview (Teachers and Nomads), and the Thrifty worldview (Gatherers and Warriors).

# Suggested Reading List and Going Further

## Learning More About GenoType

For those who wish to learn more about many of the topics and concepts behind the GenoType Diet, we've prepared a special reading list.

### Blood Types

The best resources for learning more about blood types, their role in disease, and their use in determining optimum diet are Dr. D'Adamo's extensive Blood Type Diet Library, in particular *Eat Right for Your Type* and the *Complete Blood Type Encyclopedia* (both published by G.P. Putnam Sons, New York City). Or visit the Web site www.dadamo.com.

### Secretor Status

The only book in print that currently discusses the role of secretor status in health and disease is *Live Right for Your Type* (published by G.P. Putnam Sons, New York City).

## Index-to-Finger Ratio (D2:D4 ratio)

The best book on the subject of digit ratios is *Digit Ratio: A Pointer to Fertility, Behavior and Health* (Rutgers University Press, 2002) by John T. Manning, who is a leading investigator in the field.

## Somatotyping

The field of somatotyping has not seen many new works in recent years. William Sheldon's *Varieties of Human Physique*, written in 1940, is still the hallmark publication; it is sometimes found in libraries and is available on Internet auction sites like www.ebay.com. Sheldon's massive tome *The Atlas of Man*, published in 1954, contains thousands of photographs of each somatotype. A more recent book, *Somatotyping—Development and Applications* by J. E. Carter and Barbara H. Heath, is a bit technical but does do a good job of explaining the objective ways of determining somatotype.

## Genetics

There are many good introductory books on genetics. In particular, those by Matt Ridley such as *The Agile Gene* and *Genome* are highly recommended. Both are published by Harper Perennial. A thought-provoking examination of the role of religion in genetics can be found in *The Language of God* by Francis S. Collins, published by Free Press. For the other side of the coin, try Richard Dawkin's masterpiece, *The Selfish Gene* (Oxford University Press). At a whole other level, Tara Rodden Robinson's *Genetics for Dummies* (Wiley), is simple and straightforward.

Some additional books you may find interesting are *The Triple Helix* by Richard Lewontin (Harvard University Press), *The Dependent Gene* by David S. Moore (Henry Holt and Co.), *The Elegant Universe* by Brian Greene (Vintage Books), and *Genes in Conflict* by Austin Burt and Robert Trivers (Harvard University Press).

## Dermatoglyphics (Fingerprint Analysis)

There are very few published books on the scientific significance of fingerprint analysis. *Fingerprints, Palms and Soles* by Harold Cummins is the premier work on the subject but is long out of print. It can sometimes be found in libraries or available on Internet auction sites like www.ebay.com. *Dermatoglyphics in Medical Practice* by Blanka Schaumann and Milton Alter, published in 1976 by Springer-Verlag, is another good resource.

## Haplogroups and Ancestral DNA

Two very good books on the subject of mitochondrial and Y chromosome DNA ancestry are *The Journey of Man* by Spencer Wells (Random House Trade Paperbacks) and *The Seven Daughters of Eve* by Brian Sykes (W. W. Norton & Company). Two other important works on the subject are *The Great Human Diasporas* by Luigi Luca Cavalli-Sforza and Francesco Cavalli-Sforza (Basic Books) and the *History and Geography of Human Genes* by Luigi Luca Cavalli-Sforza, Paulo Menozzi, and Alberto Piazza (Princeton). Good information on ancestral DNA and test kits are available at the National Geographic Genographic Project Web site: www.nationalgeographic.com/genographic/.

## Nutritional Genomics (Nutrigenomics)

The field of nutritional genomics is still in an embryonic stage and has yet to see the publication of a cogent, scholarly, yet accessible work on the subject. For those really interested in the field, *Nutritional Genomics*, edited by Jim Kaput and Raymond L. Rodriguez (Wiley-Interscience), and *Nutritional Genomics,* edited by Regina Brigelius-Flohe and Hans Georg Joost (Wiley-VCH), are recommended.

## Biometrics (Body Measurements)

A good guide to learning more about performing physical measurements is *Handbook of Physical Measurements* by Judith Hall, Judith Allanson, Karen Gripp, and Anne Slavotinek, published by Oxford University Press, USA. This book even contains some information on fingerprint and palm measurements. However, reference to the gonial (jaw) angle and the upper-leg interspace measurement and their clinical significance can be found only in the ultra-rare *Human Constitution in Clinical Medicine* by George Draper, C. W. Dupertuis, and J. L. Caughey, Jr. (published by Paul B. Hoeber, Inc. New York, 1944).

**Epigenetics and the Prenatal Influences on Later Life**

Epigenetics is a new and fast-evolving science, and no doubt the best books in the field are yet to be written. However, *Evolution in Four Dimensions: Genetic, Epigenetic, Behavioral, and Symbolic Variation in the History of Life* by Eva Jablonka, and Marion J. Lamb (MIT Press) is very good. *The Fetal Matrix: Evolution, Development and Disease* by Peter Gluckman, Mark Hanson (Cambridge University Press) is an excellent overview of the fetal environment's effects on health, disease, and mortality.

# GenoType Lifestyle Support

As you embark on your GenoType exploration, there are many useful resources that will make your transition easier and connect you to a community of like-minded individuals.

**www.genotypediet.com**

Please visit the official GenoType Diet Web site at www.genotypediet. com for recipes, meal plans, bulletin boards, GenoType Calculators, and tools that have been specially designed to support your GenoType journey. This Web site is designed to assist you in the process of living at your full genetic potential. Also available on the site are:

- The GenoType Diet Custom-formulated supplements designed by Dr. D'Adamo to support optimal health for each of the six Geno-Types.
- The GenoType Diet Biometric Measurement Kit, which contains all the essential tools you'll need to complete your measurements at home: a single-use inked fingerprint film, fingerprint cards, magnifying glass, metric ruler, protractor, taster strips, and tape measure.
- The At Home Blood Typing Kit, which allows you to determine your blood type in five minutes at home.
- The Secretor Submission Test kit, which allows you to determine your secretor status. This is a mail-in saliva sample submission kit, and you'll typically receive your results in about three weeks.

**Dr. D'Adamo's next book is already on your computer.**

Introducing Swami Xpress™, the companion software for The GenoType Diet.

Now that you've read the book, you can take your journey to personalized nutrition to another level by utilizing the new dietary software program written by Dr. Peter D'Adamo.

Dr. D'Adamo has combined the power of computers with his encyclopedic knowledge of biochemical individuality to produce the ultimate diet software program. Swami Xpress is simple-to-use software that allows you to generate a unique, one of a kind dietary report that is matched to the results of your personal GenoType Diet data. With Swami Xpress you will get your very own food lists and recipes, culled from hundreds of recipes designed specifically for The GenoType Diet. You will also get access to a unique meal planner, which helps you create individualized weekly meal plans and shopping lists, as well as cookbooks that contain recipes that are right for you!

For more information about Swami Xpress™ please contact:
D'Adamo Personalized Nutrition
North American Pharmacal, Inc.
213 Danbury Road
Wilton, CT 06897
(203) 761-0042
Toll Free: 1-877-226-8973
www.RightForYourType.com

## GenoType Practitioner Support

For health professionals who are interested in utilizing the concept of the GenoType Diet and the Blood Type Diet in clinical practice, there are numerous resources for products, test kits, education, and certification.

### Institute for Human Individuality (IfHI)
The Institute for Human Individuality (IfHI), a 501(c)3 under Southwest College of Naturopathic Medicine, has as its prime goal the fostering of

education and research in the expanding areas of epigenetics and bio-chemical individuality.
IfHI Educational Services, LLC
213 Danbury Road
Wilton, CT 06897
(203) 761-0042
www.ifhi-online.org

### The New England Center for Personalized Medicine
Under the supervision of Dr. Peter D'Adamo, the New England Center for Personalized Medicine employs the principles of the GenoType Diet and the Blood Type Diet as part of a complete-person approach to health. For information on appointments, contact:
The New England Center for Personalized Medicine
213 Danbury Road
Wilton, CT 06897
(203) 834-7500
www.dadamo.com/clinic

## Blood Type Diet Resources

To find out more about Dr. D'Adamo's earlier works on blood types and related products, contact:
North American Pharmacal, Inc.
213 Danbury Road
Wilton, CT 06897
203-761-0042
Toll Free: 1-877-226-8973
www.RightForYourType.com

# Index

## A

ABO blood types, *see* Blood types
Acetylation, 41
  in Explorer GenoType, 153–54
Acetyl groups, 20
Adrenaline, 122
Advanced GenoType Calculator,
    38–40, 56, 63–64
  use of, 95–97
Advanced gylcation end products
    (AGE), 26–27
Aging
  epigenetics and, 25–27
  of Nomad GenoType, 173
  of Teacher GenoType, 141
Agouti mouse, 22–23
Alleles, 3–4
  methylation and, 18–20
Allergies, among Explorer
    GenoType, 149–51

Alzheimer's syndrome
  in family history, 67
  fingerprints correlated with, 44
  genetic testing for, 28–29
Ambidextrousness, 74
Amino acids, 186
Andric body types, 46–47
  Hunter GenoType as, 122
Androgens, 61
Anemia, among Explorer
    GenoType, 153
Arches, in fingerprints, 71
Asymmetries
  among Explorer GenoType, 153
  among Gatherer GenoType, 132
  among Nomad GenoType, 174
  of fingers, 45–46
Autoimmune disease, in family
    history, 68
Ayurveda, 111

# B

Baltimore, David, 16
Barker, David, 24
Basic GenoType Calculator, 33–36,
  56, 57
  index- and ring-finger lengths for,
    61–62
  Intermediate GenoType
    Calculator compared with, 95
  leg and torso length measurements
    for, 58–60
  use of, 85–89
Beverages, 188
  in Explorer GenoType diet, 248
  in Gatherer GenoType diet, 216
  in Hunter GenoType diet, 201
  in Nomad GenoType diet, 281
  in Teacher GenoType diet,
    231–32
  in Warrior GenoType diet,
    264–65
Biometrics, 46–48, 74
Birth, 6
Blood types
  in choice of GenoType
    Calculator, 33
  of Explorer GenoType, 153
  in Intermediate GenoType
    Calculator, 37–38, 62, 90–95
  testing for, 63
Body mass index (BMI), 49
Body types, 47–48, 76–77
Brachycephalic head shape, 83

# C

Caffeine sensitivity, 41, 67
  among Explorer GenoType, 152
Cancer
  among Teacher GenoType, 143

in family history, 68
Carabelli's cusp, 48, 78
Carbohydrates, 187
  in Explorer GenoType diet,
    243–44
  in Gatherer GenoType diet, 212
  in Hunter GenoType diet, 197–98
  in Nomad GenoType diet, 276–77
  in Teacher GenoType diet, 228
  in Warrior GenoType diet,
    260–61
Cardiovascular system, of Warrior
  GenoType, 162
Celiac disease, 72, 73
Cephalic index, 81–83
Cheeses, 185
Chemical sensitivity, among
  Explorer GenoType, 154
Chronic fatigue syndrome, 170,
  171
Circulatory system, of Warrior
  GenoType, 162
Composite fingerprints, 72
Condiments and additives, 189
  in Explorer GenoType diet,
    248–49
  in Gatherer GenoType diet,
    216–17
  in Hunter GenoType diet, 201–2
  in Nomad GenoType diet, 281–82
  in Teacher GenoType diet, 232–33
  in Warrior GenoType diet, 265
Corrigan, Douglas ("Wrong-Way
  Corrigan"), 152

# D

Dairy products, 185
  in Explorer GenoType, 240–41
  in Gatherer GenoType diet, 208–9

in Hunter GenoType diet, 194–95
in Nomad GenoType diet, 273–74
in Teacher GenoType diet, 225–26
in Warrior GenoType diet,
257–58
Deltas, in fingerprints, 70, 72
Depression, in family history, 67
Dermatoglyphics, 43
Detoxification, 219, 252
Diabetes, in family history, 68
Diet
for Explorer GenoType, 154–56
for Gatherer GenoType, 134–36
for Hunter GenoType, 124–26
for Nomad GenoType, 175–78
prenatal, 5
for Teacher GenoType, 143–46
"Ten Commandments" of
successful lifestyle, 181–83
for Warrior GenoType, 165–67
Digestion, fingerprints tied to,
72–73
Diseases, caused by gene
interactions, 29
Dolichocephalic head shape, 82
Dominant alleles, 3–4
Dry brushing, 284–85

E

Ectomorph body type, 47, 48, 77
Teacher GenoType, 142
Eggs, 184
in Explorer GenoType, 240
in Gatherer GenoType diet, 208
in Hunter GenoType diet, 194
in Nomad GenoType diet, 273
in Teacher GenoType diet, 225
in Warrior GenoType diet, 257
Endomorph body type, 47, 48, 77

of Gatherer GenoType, 132
Environment
epigenetics and, 16–17, 43
prenatal, 5–6, 21–24
worldviews of responses to,
113–14
Epigenetics, 15–18, 43, 104, 287,
288
aging and, 25–27
inheritance of, 21–22
resources on, 306
Estrogen, 46, 61
Ethnicity, 8–9
Evolution, 286–87
Exercise, 182, 190
for Explorer GenoType, 251–52
for Gatherer GenoType, 219–20
for Hunter GenoType, 203–4
for Nomad GenoType, 284–85
for Teacher GenoType, 234–36
for Warrior GenoType, 267–68
Explorer GenoType, 11–12,
147–56
all blood types in, 37
diet for, 154–56
GenoType Diet for, 237–52
immune system profile for,
153–54
metabolic profile for, 153
problem areas for, 152
reactive worldview of, 113
table for, 101
table of features of, 148–49

F

Family histories, 67–68
Famine, 23–24
Fast acetylation, 41
of Warrior GenoType, 164

Fats and oils, 186–87
  in Explorer GenoType diet, 243
  in Gatherer GenoType diet, 211
  in Hunter GenoType diet, 196–97
  in Nomad GenoType diet, 275–76
  in Teacher GenoType diet, 227
  in Warrior GenoType diet, 260
Fingerprints, 43–45
  of Explorer GenoType, 153
  interpreting, 70–72
  of Nomad GenoType, 171, 174
  symmetries of, 46, 73–74
  taking, 69–70
  of Teacher GenoType, 142–43
  of Warrior GenoType, 164
  white lines in, 72–73
Fingers
  measuring lengths of, 35–36,
    61–62
Fish and seafood, 185
  in Explorer GenoType diet, 239
  in Gatherer GenoType diet, 207–8
  in Hunter GenoType diet, 193
  in Nomad GenoType diet, 271–72
  in Teacher GenoType diet, 223–24
  in Warrior GenoType diet,
    255–56
Foods
  ability to taste, 51–52, 83–84
  categories of, 183–89
  "Ten Commandments" of
    successful lifestyle, 181–83
  *see also* Diet; Superfoods
Fruits, 188
  in Explorer GenoType diet,
    246–47
  in Gatherer GenoType diet,
    214–15
  in Hunter GenoType diet, 199–200
  in Nomad GenoType diet, 279–80

  in Teacher GenoType diet, 230
  in Warrior GenoType diet, 263


**G**

G6PD enzyme, 153
Gatherer GenoType, 10, 127–36
  diet for, 134–36
  GenoType Diet for, 205–20
  immune profile for, 133
  metabolism of, 132–33
  problem areas for, 131–32
  table for, 99
  table of features of, 128–29
  thrifty worldview of, 113–14
Genes
  alleles, 3–4
  disease caused by interactions
    among, 29
  epigenetics and, 15–18, 43
  methylation and, 18–20
  mutations of, 287
Genetics
  evolution and, 287
Genetic testing, 28–29
GenoType Calculators, 29–32, 56,
  57
  Advanced, 38–40, 63–64, 95–97
  Basic, 33–36, 57–62, 85–89
  Intermediate, 36–38, 62–63,
    89–95
  interpreting results of tables, 104
  strength-testing tables for,
    97–103
GenoType Diet
  for Explorer GenoType, 237–52
  food categories for, 183–89
  for Gatherer GenoType, 205–20
  for Hunter GenoType, 191–204
  for Nomad GenoType, 269–85

for Teacher GenoType, 221–36
"Ten Commandments" for, 181–83
for Warrior GenoType, 253–68
GenoTypes, 6–8
  ethnicity and, 8–9
  list of, 9–13
  practitioner support for, 307–8
  strength-testing, 40–43, 56, 65–67, 97–104
Global methylation, 19–20
Glutathione, 153
Gluten, 125, 177
  celiac disease and, 72, 73
Glycation, 26, 301
Gonial angle, 50, 79
Green tea, 19–20
Growth factors, 132
Gynic body types, 46–47
  Gatherer GenoType as, 133

**H**

Handedness, 46, 74
Haplogroups, 305
Head shape, 50, 81–83
  jaw angles, 78–79
  of Warrior GenoType, 163–64
Heart disease, in family history, 68
Height, 58–59
High blood pressure, 160
High-glycemic foods, 135
Hips, in waist-to-hip ratios, 49, 79–80
Histone acetylation, 20–21
  aging and, 25
Histones, 20–21
Hox genes, 35
Human Genome Project, 15–16
Hunter GenoType, 9–10, 116–26

diet for, 124–26
GenoType Diet for, 191–204
immune system profile for, 123
metabolic profile for, 122–23
problem areas for, 121–22
reactive worldview of, 113
table for, 98
table of features of, 117–18
Warrior GenoType compared with, 165
Hypoglycemia, 131, 177
Hypothyroidism, 132

**I**

Identical twins, 43–44
Immune system, macrophages in, 171
Immune system profiles
  for Explorer GenoType, 153–54
  for Gatherer GenoType, 133
  for Hunter GenoType, 10, 123
  for Nomad GenoType, 174–75
  for Teacher GenoType, 142
  for Warrior GenoType, 164
Incisor shoveling, 48, 77–78
  among Nomad GenoType, 175
Incomplete proteins, 186
Index-finger measurements, 35–36, 61–62
Index-to-finger ratio, 304
Inflammation, 42, 123
Inheritance, of epigenetic patterns, 21–22
Insulin-like growth factors (IGF), 34
Intermediate GenoType Calculator, 36–38, 56, 62–63
  use of, 89–95

**J**

Jaw angles, 50, 78–79
Jefferson, Thomas, 120–21
Jordan, Michael, 120, 121

**K**

Kenney, Father, 286

**L**

Lectins, 301
Left-handedness, 74
  among Explorer GenoType, 153
Leg measurements
  in Basic GenoType Calculator,
    87–88
  in Intermediate GenoType
    Calculator, 91–94
  opening between legs, 75
  and torsos, 34–35, 58–60
Lifestyle planning
  for Explorer GenoType, 251–52
  for Gatherer GenoType, 218–20
  for Nomad GenoType, 283–85
  for Teacher GenoType, 234–36
  "Ten Commandments" for, 181–83
  for Warrior GenoType, 267–68
Lincoln, Abraham, 142
Live foods, *see* Vegetables
Livers, of Explorer GenoType, 152
Loops, in fingerprints, 71

**M**

Macrophages, 171
Meats, 183–84
  *see also* Red meats
Medical histories, 67–68
Meditation, 268

Mediterranean diet, 165
Men, waist-to-hip ratios for, 49
Mesocephalic head shape, 83
Meso-ectomorph body type, of
    Teacher GenoType, 142
Mesomorph body type, 47, 77
  definition of, 301
  of Explorer GenoType, 153
  of Nomad GenoType, 173
Metabolic profiles
  for Explorer GenoType, 153
  for Gatherer GenoType, 131–33
  for Hunter GenoType, 122–24
  for Nomad GenoType, 173–74
  for Teacher GenoType, 142–43
  for Warrior GenoType, 163–64,
    167
Metabolism, 48
Methylation, 4, 18–20
  aging and, 25–26
  of histones, 21
Mini strokes, 160
Misogi, 235
Monroe, Marilyn, 132
Morphology, 74
Mutations, 287

**N**

Natural selection, 287
Neurodegenerative diseases, 173
Nitric oxide (NO), 171–73, 176
Nomad GenoType, 13, 168–78
  diet for, 175–78
  GenoType Diet for, 269–85
  immune system profile for, 174–75
  metabolic profile for, 173–74
  problem areas for, 173
  red hair and, 9
  table for, 103

table of features of, 169–70
thrifty worldview of, 113–14
tolerant worldview of, 114
Non-secretors, 40
Non-tasters, 51, 83
Nutritional genomics, 305
Nuts and seeds, 186

## O

Obesity, in Gatherer GenoType, 10
Oils, *see* Fats and oils

## P

Personal histories, 67–68
Phenylthiocarbamide (PTC), 51
Poultry, 184
    in Explorer GenoType diet, 238
    in Gatherer GenoType diet,
        206–7
    in Hunter GenoType diet, 192–93
    in Nomad GenoType diet, 270–71
    in Teacher GenoType diet, 223
    in Warrior GenoType diet, 255
Pregnancy
    prenatal environment during, 5–6,
        21–24
    starvation during, 23–24
Prenatal environment, 5–6, 21–24
    development of fingerprints in,
        43–44
    finger development in, 45–46
    handedness and, 46
    of Nomad GenoType, 174
    reading clues to, 29–30
Presley, Elvis, 133
Profiles of GenoTypes, 110–11
PROP (propylthiouracil), 51, 83–84
Protein, from vegetables, 186

## R

Reactive worldview, 113
Recessive alleles, 3–4
Red meats, 183–84
    in Explorer GenoType diet, 238
    in Gatherer GenoType diet, 206
    in Hunter GenoType diet, 192
    in Nomad GenoType diet, 270
    in Teacher GenoType diet, 222
    in Warrior GenoType diet, 254
Religion, 286–87
Rh blood types, 56–57, 63
    in Advanced GenoType
        Calculator, 95–97
    in Explorer GenoType, 151, 153
    in Teacher GenoType, 142
Ridges, in fingerprints, 72–73
Ring finger measurements, 35–36,
    61–62

## S

Saunas, 219
Seafood, 185
    *see also* Fish and seafood
Secretor status, 38–40
    testing for, 39, 63–64
Sex hormones, 35, 46
Sheldon, William, 301
Sitting height, 58
Sleep habits, 182
Slow acetylation, 41, 152
Smoking, 159–60
Somatotypes, 76
Spices, 188
    in Explorer GenoType diet, 247
    in Gatherer GenoType diet, 215
    in Hunter GenoType diet, 200
    in Nomad GenoType diet, 280
    in Teacher GenoType diet, 231

in Warrior GenoType diet, 264
Standing height, 58
Starvation, 23–24
Strength-testing GenoTypes, 65–67
  questions for, 40–43
  tables for, 97–104
Stress, 182
  for Hunter GenoType, 121–22,
    204
Strokes
  in family history, 68
  mini strokes, 160
Superfoods
  for Explorer GenoType, 155,
    237–49
  for Gatherer GenoType, 134–35,
    205–17
  for Hunter GenoType, 124–25,
    191–202
  for Nomad GenoType, 176,
    269–82
  for Teacher GenoType, 144–45
  for Warrior GenoType, 166,
    253–65
Super-tasters, 51, 84
Supplements, 26, 189–90
  for Explorer GenoType, 250
  for Gatherer GenoType, 217–18
  for Hunter GenoType, 202–3
  for Nomad GenoType, 282–83
  for Teacher GenoType, 221–34,
    233–34
  for Warrior GenoType, 266
Sweating, 251
Symmetries
  in fingerprints, 73–74
  of fingers, 45–46
Symmetries among, 45–46
Syndrome X, 132

T
Taster status, 51–52, 83–84
Teacher GenoType, 11, 137–46
  diet for, 143–46
  GenoType Diet for, 221–36
  immune system profile for, 143
  metabolic profile for, 142–43
  Nomad GenoType compared
    with, 174, 175
  problem areas for, 141–42
  table for, 100
  table of features of, 138–39
  tolerant worldview of, 114
Teeth, 48, 77–78
  of Nomad GenoType, 175
Telomeres, 25–26
"Ten Commandments" of successful
    lifestyle, 181–83
Tendons, 76
Testosterone, 46
Thrifty worldview, 113–14
Tolerant worldview, 114
  of Teacher GenoType, 140, 141
Torso and leg measurements, 34–35
  in Basic GenoType Calculator,
    87–88
  in Intermediate GenoType
    Calculator, 58–60, 91–94
Toxins
  in Explorer GenoType diet,
    237–49
  in Gatherer GenoType diet,
    205–17
  in Hunter GenoType diet,
    191–202
  in Nomad GenoType diet, 269–82
  in Teacher GenoType diet, 221–34
  in Warrior GenoType diet,
    253–65

Type AB blood, 37
  among Warrior GenoType, 163
Type A blood, 37
  among Warrior GenoType, 163
Type B blood, 37
Type O blood, 37

**V**

Vegetable proteins, 186
  in Explorer GenoType, 241–42
  in Gatherer GenoType diet,
    210–11
  in Hunter GenoType diet, 195–96
  in Nomad GenoType diet, 274–75
  in Teacher GenoType diet,
    226–27
  in Warrior GenoType diet,
    258–59
Vegetables
  in Explorer GenoType diet,
    244–46
  in Gatherer GenoType diet,
    213–14
  in Hunter GenoType diet, 198–99
  in Nomad GenoType diet, 277–78
  in Teacher GenoType diet, 228–29

  in Warrior GenoType diet,
    261–62
Visualization exercises, 172

**W**

Waist-to-hip ratios, 49
  measuring, 79–80
Warrior GenoType, 12–13, 157–67
  diet for, 165–67
  GenoType Diet for, 253–68
  immune profile for, 164
  metabolic profile for, 163–64
  problem areas for, 162–63
  table for, 102
  table of features of, 158–59
Weight, of Warrior GenoType, 162
White lines, in fingerprints, 72–73,
    302
Whorls, in fingerprints, 71
Women, waist-to-hip ratios for, 49
Worldviews associated with
    GenoTypes, 112–15

**X**

Xenobiotics, 133